FUZZY LOGIC / DANIEL MCNEILL,
TJ213 .M353 1993

D1014704

DATE DUE

MY 13 '94			
AG 4 '94			
12-19-97			
MR 29 '99			
MY 17 '99			

DEMCO 38-296

FUZZY LOGIC

■

DANIEL McNEILL

PAUL FREIBERGER

SIMON & SCHUSTER

NEW YORK LONDON

TORONTO SYDNEY

TOKYO SINGAPORE

SIMON & SCHUSTER
ROCKEFELLER CENTER
1230 AVENUE OF THE AMERICAS
NEW YORK, NEW YORK 10020

LIBRARY OF CONGRESS CATALOGING IN PUBLICATION DATA
McNeill, Dan.
Fuzzy logic / Daniel McNeill and Paul Freiberger.
p.   cm.
Includes index.
1. Automatic control.   2. Fuzzy systems.   3. Computer industry—
United States.   4. Technology transfer—United States.
5. Competition, International.   I. Freiberger, Paul.
II. Title.
TJ213.M353   1993
628.9—dc20   92-42631
CIP

ISBN: 0-671-73843-7

To Wendy O'Sullivan and Robin Said

—D.M.

To Jeanne and Edan

—P.F.

CONTENTS

■

CONTENTS

PROLOGUE

A TRAIN RUNS IN SENDAI

■

We didn't even know fuzzy logic
was important and the Japanese
had been using it for years.

—FEDERICO FAGGIN

He that will not apply new rem-
edies must expect new evils; for
time is the greatest innovator.

—FRANCIS BACON

Lotfi Zadeh stood in the cab of the shiny new subway as it flowed out of the station. Streets, trees, and rooftops drifted by below, and he recalled the classic passenger's illusion: A train starts up with gossamer ease, no jolts or tug of inertia, and the windows of the adjacent car slip behind faster and faster until one realizes that actually the other car is moving. Of course. No real train could begin so smoothly.

But this one had. As it purred toward the outskirts of Sendai, Japan, few passengers in the cars bothered to hang on to the straps, even when it stopped or started. It was also halting with uncanny precision and, he knew, saving 10 percent on fuel. Zadeh ("ZAH-da"), a professor at UC Berkeley, was not an expert on subways, but he knew this one's secret very well, for it was fuzzy logic, which he had invented in the United States.

By 1991, the Sendai subway had carried passengers for four years and was still the most advanced system on earth. In that year the city of Los Angeles was building a subway and needed a control system. It did not even consider fuzzy logic. "I've heard of fuzzy logic very vaguely and I'm not that familiar with it," said Robert Townley, engineering manager for automatic train control for this project. He added that none of the bids to design the subway's control used fuzzy logic. "We're not requiring it and the proposers have not chosen to do so."

Why not?

The simple answer is that fuzzy logic is an outcast in the United States. It is one more technology the United States has created and neglected, only to watch the Japanese pick up, nurture into profitability, and sell back to us.

This giveaway, however, was especially arresting. Some U.S. innovations, like the flat-panel computer display, required extensive further work. Others, like the VCR, fell victim to shoddy marketing and manufacturing. Still others, like the industrial robot, simply failed to spur investor interest. In all these cases, U.S. firms saw the value of the goal, but were either unable to reach it in time or felt it would not pay for itself. The blocks were technological and financial.

With fuzzy logic, the block was mental. As early as 1973, Zadeh had shown how fuzzy logic could make smarter machines, and by 1979 a Danish firm was using it in a complex industrial process and saving tens of thousands of dollars. The technique existed more or less complete. It needed no miniaturization, no years of refinement, no daring gamble of cash. Alert Japanese observed this success and pursued it, yet not a single U.S. company did the same.

The United States actually spurned fuzzy logic. Respected scholars still warn against it, calling it a sham and a fad. It tolerates unheard-of permissiveness, they say, and threatens the scientific method. Their opinions seem authoritative and final. "Delete all the material on 'fuzzy sets'—it has been proved to lack a sound mathematical basis," said one referee of a proposal before the National Science Foundation in 1989.[1]

Such indictments made fuzzy logic an underground technology in the United States. Young professors might risk their careers by pursuing it, and its meager funding attracted only the hardiest scientists, the saguaros of academia. Indeed, the animus against it ran so deep that when U.S. firms such as Motorola, Otis Elevator, General Electric, Pacific Gas & Electric, and Ford finally began venturing into the field, they found clear explanations of it wanting.

Fuzzy logic inhabits a different world in Japan. The Japanese not only welcomed fuzzy logic, but dizzily embraced it. There, instead of a trickle of information, one finds gullywashers. The nightly news routinely shows it balancing poles or controlling agile robot hands. Prime-time TV ads extol it and panels of show-biz celebrities discuss it before the nation's eyes. Even children in the street talk of it, and if the L.A. subway control manager lived in Osaka or Tokyo, he would get to know *faaji* very quickly.[2]

He would also see the gifts fuzzy logic has brought to Japan: intelligent washing machines, microwaves, cameras, camcorders, automobiles. Toss a load of clothes into a fuzzy washer and press *Start*, and the machine begins to churn, automatically choosing the best cycle. Place chili, potatoes, or lasagna in a fuzzy microwave and push a single button, and it cooks for the right time at the proper temperature. Aim a fuzzy camcorder at a birthday party and it stills handheld jitter. The Japanese sell fuzzy TVs, which ironically show sharper pictures. Fuzzy logic is making over cars: cushioning their ride, enhancing safety, and cutting gas consumption by some 15 percent. It is scheduling elevators and traffic lights, and preventing tunnel cave-ins at construction sites. Because of fuzzy logic, the visiting Westerner sees a nation living slightly further in the future.

These boons all go back to 1964, the year Zadeh invented and christened fuzzy logic. It is a charming name and a wild misnomer. Fuzzy logic is not logic that is fuzzy, but logic that describes and tames fuzziness. And even that definition falters, for most of the theory is not logic at all. It is a theory of fuzzy sets, sets that calibrate vagueness.

Fuzzy logic rests on the idea that all things admit of degrees. Temperature, distance, beauty, friendliness, greenness, pleasure—all come on a sliding scale. The Canadian Rockies are *very beautiful*. My next-door neighbor is *fairly lazy*. Boston is *quite close* to Cape Cod. Likewise, objects are objects to degrees. Astronomers say Jupiter is a *star* to a weak extent. Egypt was partly a *colony* of Britain; the United States was largely one. A dagger is very much a *weapon*, while a curtain rod is scarcely a *weapon* at all. Such sliding scales often make it impossible to distinguish members of a class from nonmembers. When does a hill become a mountain, or a pond a lake? How far is far?

Traditional logic, set theory, and philosophy have compelled sharp distinctions. They have forced us to draw lines in the sand. For instance, a *novel* might have 90 pages or more, a *novella* less than 90. By this standard, a 91-page work would be a novel, an 89-page one a novella. Thus, if printers reset the novella in larger type, it becomes a novel. Fuzzy logic avoids such absurdities.

Its future could be even more brilliant, for fuzzy logic reflects how people think. It partly models our sense of words, our decision making, our recognition of sights and sounds. It unveils a corner of intuition. It may also mirror actual brain functions, such as detecting color and distinguishing phonemes. As a result, it is leading to new, more human machines.

For instance, computer scientists have toiled for years on expert systems, software that makes decisions like a doctor, geologist, or financial analyst. An expert system for diagnosing illness might identify ailments better and faster than the top internists, and save myriad lives. Despite some successes, however, expert systems have hit a wall. Fuzzy expert systems could crack through it. "I think we'll begin to see a lot of fuzzy expert systems," says K. K. Yawata, president of LSI Logic in Silicon Valley. "I think there'll be another wave after this consumer applications tide."

Fuzzy logic has led to new devices that identify written characters and speech. Past computers have required a near-perfect match between a letter and a pattern in memory. But letters come in a kaleidoscope of fonts, and handwriting is infinitely various. Sony now markets an inexpensive fuzzy computer that recognizes handwriting in English and Japanese, and Ricoh is developing one that identifies spoken words.

As the effort in artificial intelligence (AI) groaned to a crawl in the mid-1980s, snarled in its own complexity, a new kind of computer appeared: the neural network. These devices learn. Based crudely on the brain, they offered soaring promise at first, but they too have stalled. Linked to fuzzy systems, however, they become more powerful and reliable. Such devices could jumpstart AI.

Zadeh describes a true machine of the future: a palm-sized personal secretary. "You talk to it and you say the sorts of things you might say to a human secretary, like, 'I filed something in a certain location,' " he says. To find this document, the little box would face an array of now-perplexing tasks: recognizing the words, understanding them, and recalling imprecise information. "Without fuzzy logic, you can't solve those sorts of problems," Zadeh says.

How far can fuzzy logic go? In a few decades, says Bart Kosko, a professor at the University of Southern California, it will lead to marvels:

- Vast expert decision makers, theoretically able to distill the wisdom of every document ever written.
- Smart cars with sonar devices that pump the brakes if the car ahead on the freeway stops too quickly. With a fuzzy navigator, a computerized map, and emitters and receivers in the asphalt, such vehicles could drive themselves.
- Sex robots with a humanlike repertoire of behavior.
- Computers that understand and respond to normal human language.
- Machines that write interesting novels and screenplays in a selected style, such as Hemingway's.
- Molecule-sized soldiers of health that will roam the bloodstream, killing cancer cells and slowing the aging process.

Should such devices come about, who will make and batten off them? Probably the Japanese. Lotfi Zadeh estimates that the United States is now some five to six years behind Japan, and the distance is growing every day. In well-funded laboratories across the country, the Japanese are pushing the horizon of fuzzy logic in all directions. Moreover, their aggressive pursuit of fuzzy appliances has poised them to swallow up new U.S. markets. Their one-button washing machines and microwaves have obvious consumer appeal, as do their gas-saving cars.

How did it happen? How did the United States blind itself to a commercial jackpot?

1

THE MASTER

The things of the universe are not
sliced off from one another with
a hatchet, neither the hot from the
cold nor the cold from the hot.

—ANAXAGORAS

There are no whole truths; all
truths are half-truths. It is trying to
treat them as whole truths that
plays the devil.

—ALFRED NORTH
WHITEHEAD

You have seen him spout; then
declare what the spout is; can
you not tell water from air? My
dear sir, in this world it is not so
easy to settle these plain things.

—HERMAN MELVILLE,
MOBY-DICK

■

If it weren't for a broken dinner engagement, fuzzy logic might never have come into being. In July 1964 Lotfi Zadeh was staying at his parents' apartment in New York and planned to dine out with a friend. At the last moment the friend had to cancel and a free evening loomed up. Zadeh considered what to do. He normally keeps a list of problems in the back of his mind and periodically selects one to ponder. Moreover, he had promised to work at the Rand Corporation later that month and had not yet chosen a research topic. So he lay down on a bed, his preferred posture for cogitation, and contemplated complex systems. And the notion of fuzzy sets struck him.

"It was a ridiculously simple idea," he says, "but I could see its importance immediately." Indeed, it blazed out so clearly he thought of writing a note predicting it would be one of the most pivotal concepts ever to emerge from his department at Berkeley, putting it in an envelope, and mailing it to imprint a date on the envelope. He refrained, viewing the plan as immodest.

Zadeh also realized that he could unfurl its implications very quickly, and after half an hour he had worked out its basic contours. He continued to lie there thinking, writing nothing down, and three hours later, when he was ready to fall asleep, he had "a reasonably good idea of how this whole thing would look."

A slight figure with a bald head, wide in back and tapering down to a pointed chin, Zadeh is a courtly gentleman in the Old World style, and the fuzzy community abounds with tales of his generosity, encouragement, and insight. "He has the ability to know people very quickly, sometimes better than they know themselves," says John Yen, a former student. Even among professional adversaries, few have harsh words for him and many note his tact and intelligence. And friends instantly recognize his signature line: "Of course, so-and-so [an antagonist] and I are very good friends."

He does not shun the unconventional. In his otherwise elegant home in the Berkeley hills, he has placed great black banks of stereo speakers, 28 in all. His narrow closetlike office is famous for its peculiarity— shelves to the ceiling, trays awash with papers, folders stacked beneath chairs, books in half the tongues from Krakow to Kyoto. "His memory is prodigious for names and facts," says his wife, Fay. "But he can be absentminded. Once he couldn't find a sock, and I said, 'You have both of them on one foot.' "

Though unfailingly pleasant and polite, Zadeh remains somewhat enigmatic. "He is very difficult to know," says Fay. "He asks questions of other people, but if you try to pin him down, it's very hard." Fuzzy theorist Ron Yager adds, "Lotfi is a very, very private man. He is open, but private also. What's always amazed me is Lotfi is a very *precise* person, and I've always wondered why he's involved with this topic."

That involvement began in the early 1960s, when Zadeh became enmeshed in a major scientific problem: complex systems. Equations can model simple systems like a pendulum, and statistics can describe huge disorganized systems like gas molecules in a jar. But math staggers with biological or humanistic systems. "The villain, of course," Zadeh later wrote, "is the tremendous complexity of living organisms— complexity which is orders of magnitude greater than that of the most complex inanimate systems made or observed by man."[1]

Complex systems defy human comprehension and evade even definition. They lie somewhere between the simple order of a quartz crystal and the utter disorder of the air around it, and in fact show a higher level of organization than either. That level need not be lofty. A handful of rules acting on a small motley of items can create a complex system like chess. One cannot determine the best opening in chess. It is just too complicated. A computer the size of the earth, working at top theoretical speed for the 4 billion years of terrestrial history, would not identify all possible games.[2] We can only know that some openings are better than others.

Complexity reaches its apex with life and society. Biological and social systems can be marbled with subsystems and sub-subsystems and sub-sub-subsystems, all of unthinkable intricacy. For instance, the economy is a complex system. It reacts to politics, weather, new technology, government decisions, pure emotions like panic, and much else besides. Each in turn reflects an extremely complex system. What determined yesterday's coffee futures? What caused the Depression? How have advances in biotechnology affected the U.S. balance of trade? No one knows, precisely, and no one ever will.

The brain is a complex system. Language is a complex system. High society is a complex system. The spread of disease is a complex system. Law is a complex system. A violet, a marine estuary, a political party—all are complex systems. Complexity dominates the life, behavioral, social, and environmental sciences, as well as medicine and some technology. Such systems often behave unpredictably, and they interest us much more than pendulums or air in a jar. As early as 1948, scientist Warren Weaver said science had to learn to deal with them.[3]

But how?

Zadeh had always believed in precision, yet the more he examined complexity, the more he wondered. As he co-wrote a book on linear system theory with Charles Desoer, he realized that many systems defy exact description by their very nature. He found reasonable precision in linear systems (like a spring, where the variables grow in regular proportion), stable systems (like a pendulum, which return to equilibrium after a disturbance), and time-invariant systems (like a table, which change little in the short run). Yet others thwart exactitude. They include decentralized systems (like economies, in which subunits have some control over themselves) and slowly-varying systems (like an aging person or a moving glacier).

"This was something that was bothering me," he says.

As early as 1962, he wrote that to handle biological systems "we need a radically different kind of mathematics, the mathematics of fuzzy or cloudy quantities which are not describable in terms of probability distributions."[4] It was the debut of *fuzzy*.

A month after the first flash, he had formalized the idea and presented it to his friend Richard Bellman (1920–1984) at Rand. Bellman was the brilliant, unorthodox mathematician who invented dynamic programming, whose importance in computing Zadeh likens to that of penicillin in medicine. He responded enthusiastically, greatly heartening Zadeh. Zadeh and Bellman later coauthored several papers, but Bellman did not initially help develop the theory. Rather, Zadeh says, he was "a person who cheered me on." Zadeh then wrote a paper outlining the theory of fuzzy sets. It has become one of the most referenced articles in its field.

He was born Lotfi Aliaskerzadeh in 1921 in Baku, in Soviet Azerbaijan (and changed his name to Lotfi Asker Zadeh on reaching the United States). A fast-growing town of perhaps 600,000 on the Caspian Sea,

Baku had a section called "Black City," where oil oozed up from cracks in the ground and derricks forested the half-moon bay. A beautiful palm boulevard lined the shore, where Zadeh and his friends liked to promenade past the parks and bathhouses. In summer, fierce north winds blew sand and dust through the city, forcing policemen and horses to wear goggles, but also driving the rainbow sheen away from shore, so the Zadeh family could swim in the sea.

He was the son of multiethnic, cosmopolitan parents. His Russian mother was a pediatrician, and his Turkish–Iranian father covered Baku for Iranian newspapers and engaged in import–export. Since the Soviet government coddled foreign correspondents and his father earned a good income, Zadeh recalls, "We had everything. There was alway a maid, always tutors." The family summered at spas in the Caucasus, soaking in the mineral waters and attending evening concerts. "Even though we were in the Soviet Union," he says, "I lived the life of a son of well-to-do parents in some other society."

However, the Zadehs were an island amid tumult. In 1921, the country lay prostrate from its Civil War, in which some 20 million Russians died from battles, executions, drought, famine, and illness. (In contrast, about 10 million soldiers from all nations died in World War I.) To heal and revive the country, Lenin launched the New Economic Policy (NEP), a plan of moderation which allowed small-scale capitalism. The NEP was to be the anteroom of the revolution.

Zadeh attended the first three grades of elementary school in the Soviet Union and recalls the excitement about the future, the sense that citizens were joining together to mold a new society. He saw it everywhere: in schools and newspapers, in slogans, posters, demonstrations, and parades. "You were influenced to put your own interests last and to dedicate yourself to working for society, for science." He never forgot it, and it made him sanguine about human nature.

He also imbibed the Russian love of reading. By 14, he had read "all the classics," including a Russian translation of Shakespeare, and amassed a personal library of between 2,000 and 3,000 volumes. "I was an only child, so my parents lavished a great deal of attention on me," he says.

The accession of Stalin in December 1927 ended the NEP. In October 1928 he commenced the First Five-Year Plan (which actually lasted four years and three months). It was a juggernaut. The Soviet

Union funneled huge sums into its economy, virtually creating its chemical, machine tool, aviation, automobile, and electrical industries, and causing great cities like Magnitogorsk to sprout overnight in the wilderness. The Plan also collectivized agriculture. When small land-owners resisted, 5 million of them and their families vanished. Soon famine gripped the Ukraine, and in 1931 a terrible drought began. That year Zadeh's family returned to Teheran.

Teheran was then a town of some 400,000 on a slanting plain beneath the snowy Elburz Mountains. Its ruler Shah Reza Pahlavi I (1878–1944) was also frogmarching his country into the future, and Zadeh watched with approval as he built great railroads and boulevards, erected factories, modernized the calendar, and founded the University of Teheran.[5] "Every time you turned around you saw something new," Zadeh says. "Before your eyes you saw the country transformed." Yet the Shah was also appallingly brutal. Some of Zadeh's leftist teachers at school and in the university simply disappeared. "It was a little like what happened in Argentina," he says. "There was a dreaded secret police. You didn't dare to open your mouth. And you had to stay out of trouble. The Shah became so powerful that he could do whatever he wanted. He could order someone executed right on the spot. Today you have your house and tomorrow they show up and say, 'The Shah likes your house so get out.' "

Zadeh's father continued in business, his mother set up a practice, and the family's good life continued. But Zadeh felt the culture shock nonetheless. His parents enrolled him in the American College, a Presbyterian missionary school. Zadeh attended chapel every morning, a drastic change from the atheistic Soviet schools. "I became tolerant as a result of that, having seen this fanaticism in the Soviet Union and then Presbyterianism, and also later the Muslim fanaticism," he says. "I began to realize that people can feel very strongly about things that are diametrically opposite."

He also met his future wife, Fay, who found him "interesting, unusual, fascinating. I was a little afraid of him, because he was so bright, you know." She recalls the special room Zadeh's parents set aside for him. It contained his books, and over the big table in the middle he had placed a large sign in Russian: ODIN ("alone"). She used to discuss literature with him there. But one day when he was gone, she entered this room and let out a shriek. "On the table there was a skull, a baby's skull," she says. "And in that skull there was a red light, so those red, socketless eyes were gleaming at me and the room was fairly dark, the sign ALONE, and with that skull, I thought, 'I don't want to have anything to do with this person.'"

She recalls him as highly disciplined about school, partly because he faced obstacles. Instructors taught in Persian, and Zadeh had to learn it. Today, he still speaks Russian well enough to pass for a native, but Persian words can elude him. Moreover, the curriculum shifted. Despite the new boulevards and factories, teachers steeped students in Iran's luxuriant past, and he had to memorize thousands of lyrics by Omar Khayyam (1038–1123), the wine-loving Hafez (1320–1389), and the lofty Sa'adi (c. 1190–1290). Zadeh had always been interested in technology, and the poetry irked him.

"Lotfi was always a very curious boy," says Fay, "very, very original and very respected by the kids because he invented things. He was not like them." For instance, at 13 he designed models of an independent front suspension. By 1937, when he was 16, he had received several patents, one for a rotary engine.

After eight years at the American College, he took the exam for the University of Teheran. Despite the language problem, he scored third in the nation. Newspapers announced the test results and Zadeh became well known. The public ratings continued in the university itself, where Zadeh ranked number one for the first two years. He majored in electrical engineering, avoiding a mathematics degree because in Iran it would have channeled him toward a low-ceiling career as a high school teacher.

The onset of World War II sparked excitement in Iran. "There was a great deal of hatred for the British," Zadeh says. "They were despised. And so anybody who was an enemy of your enemy was your friend. For that reason alone Germany was very popular. And I didn't feel that way, in part because I came from the Soviet Union. I was not a typical Iranian, you see."

On June 22, 1941, Hitler attacked the USSR along a front from the Baltic Sea to the Black Sea. The Shah was pro-German, but Iran instantly became the safest route for Britain and the United States to ship supplies to Soviet Union. Hence, the USSR invaded Iran in August 1941, quickly defeated the Shah's venal army, and—as it had in World War I—split Iran with Britain. Americans soon joined them, with

30,000 soldiers. The Shah fled and a welter of political parties sprang up. The turmoil emptied the department, and Zadeh was one of only three electrical engineering students to graduate in 1942.

For the next year, he worked with his father as a contractor for the U.S. Army in Iran, supplying building materials and hardware. He came to know many Americans and in 1943 decided to move to the United States. The voyage was straight out of *Casablanca*. He waited months in Iran for the necessary papers. At last he left for Cairo, where he languished several more months until a vessel appeared. Finally in July 1944, he boarded a Portuguese passenger ship bound for the United States. The crossing was tense. German submarines prowled beneath the waves and he was constantly seasick. After two weeks of agony, he debarked in Philadelphia.

He felt exhilarated, like a "fish in water." In the fall of 1944, he entered MIT, across the Charles River from Boston, as a graduate student. Even at nine or ten, he had heard of the "Massachusettsky Institute" and felt its tug. Yet he found MIT very easy, less rigorous than the University of Teheran.

The times were exciting intellectually. The Computer Age was dawning. Norbert Wiener (1894–1964) was at MIT developing cybernetics, the discipline of maintaining order in systems, which included the automatic control of machines by computers. ("I wish to describe such machines in terms which are not too foreign to the actual observables of the nervous system, and of human and animal conduct."[6]) Claude Shannon had newly minted information theory, and Warren McCulloch and Walter Pitts had sketched out neural networks, computers modeled on the brain. "It was as if we were entering a new era," Zadeh says, "in which not just computers, but information theory, communication theory, and all these theories would make it possible to create a world in which information plays a major role. There was a tremendous amount of enthusiasm and faith in the future."

He received his master's in electrical engineering in 1946. Meanwhile, his parents had settled in New York, so Zadeh applied to Columbia University and not only won entrance to its doctoral program, but a job as instructor as well. He took his Ph.D. in 1949 and next year became assistant professor at Columbia.

Thinking, and the possibility of automating it, had come to intrigue him. In January 1950 he described an electronic director of admissions—

"it was not very serious"—based on a series of IF–THEN rules.[7] It was a decision maker, a primeval expert system. He also declared the importance of multivalued logics, logics with values between true and false. The time was not far off, he stated, when students would find a course in mathematical logic just as valuable as a course in complex variables.

At MIT he had studied under Ernst Guillemin, an influential teacher whose "life was circuitry and working with students." But by then he had come to feel that Guillemin's focus was too narrow. "To me it was obvious that circuitry was mere esthetic stuff and what we needed was more general theory," he says. The larger topic, he thought, was systems. In 1954 he wrote a paper called "System Theory," which founded and christened a new discipline.

There is some confusion over this term. In 1951, before Zadeh coined *system theory*, the Hungarian biologist Ludwig von Bertalanffy (1901–1972) had expounded on "general systems theory" (with an *s*). He mingled biology and philosophy, and avoided mathematics. "There were essentially two camps," says Zadeh—himself and the electrical engineers, and Bertalanffy and the biologists. "And the camp I was a member of did not think too much of the other camp. To us those people were crackpots. They took a sort of mystical view of the thing." Their approach faded, and *system theory* today refers to Zadeh's view. It made him well known in the field, and indeed many fuzzy scientists, such as Toshiro Terano, first heard of Zadeh as a system theory man.

He earned a full professorship in 1957, capping a very rapid rise. Near the end of 1958, he received a call from John Whinnery, chairman of electrical engineering at UC Berkeley, wondering if Zadeh would like to move west. Berkeley had the best electrical engineering department in the world, but Zadeh liked Columbia and mulled the offer for weeks before accepting. He arrived at Berkeley in 1959 and became head of the department in 1963. And as the summer of 1964 edged into fall, he wrote his most famous paper.

It described a fuzzy set theory that seemed instantly full-grown. But most innovations spring from many sources, as a network of rivulets tumbles down mountains, converging into streams, until by some junction there is a Missouri or Euphrates. Zadeh's paper too brought together a fan of tributaries—set theory, the philosophy of vagueness, multivalued logic, and Max Black's word usage charts. They form the ultimate source of fuzzy theory.

DANCE OF THE SILHOUETTES

At the heart of fuzzy logic lies a question that goes to the nub of thought: What is a class?

Categories pervade our thinking. Even animals classify constantly and automatically. A cat sees a small gray creature dart across the floor, instantly spots it as a mouse, and gives chase, though it has never seen this particular mouse before. By slotting it into *mouse*, the cat can tap memory and call up information about it. Indeed, classification has proved so vital in evolution that it has passed partly beyond our control. If a rattlesnake bites a hiker, he may become wary of all snakes, though he knows few are rattlers. He can't help classifying.

Language likely arose sometime between 200,000 and 30,000 years ago. It is the supreme expression of classes, for most words refer to categories. *Street* indicates the class of street, not a particular thoroughfare. When we say *a street*, we mean one out of the class of streets. *This desk* is this particular example of the class of desks. Words and categories highlight the gist and dim down the unique details. They generalize and simplify, and without them, language would evanesce.

It is almost impossible to overstate the importance of categories. Indeed, in 1970 theorist David Marr suggested that handling classes is the paramount role of the neocortex, the gray matter of the brain.[8] It lies at the root of us all.

Mathematicians and logicians depict classes with formal models. Zadeh's fuzzy sets are such a model, and they build on the set theory which the German Georg Cantor (1845–1918) developed in the late 19th century. Cantor was a pious egocentric loud, witty, and domineering, and he loved to hold forth before a circle of admirers, shocking them with rash utterances. He was also one of the most original mathematicians in history, and developed set theory in the course of proving that infinity comes in many sizes.

Sets enabled Cantor to show that infinity plus any segment of it is also infinity. The part equaled the whole. Moreover, he demonstrated that infinity comes in different levels, an endless number, in fact. These discoveries were highly counterintuitive. For instance, he found that an inch had the same number of points as in an infinite line—or a cube, all three-dimensional space, or space of any dimension. One could draw a line from each point in infinite space to its own unique point on a one-inch line. "I see it, but I don't believe it," he wrote mathematician Richard Dedekind.[9]

Set theory is a tamer field—the pedestal to this exotic objet d'art—but one of greater usefulness. Cantor defined sets as collections of definite, distinguishable objects of our intuition or intellect. Hence, a class is a set. Sets can also represent words, since most words are classes. For instance, *a desk* is both one out of the class of desks and one out of the set of desks. Tight threads bind sets, classes, and words, and Cantor's theory affects them all.

The basics of set theory are simple, and anyone who has taken the New Math has seen them already. The New Math depicts sets as circles: lines dividing the world into in and out, true and false. This fact is revealing, for Cantor's sets are crisp. They are silhouettes. Each potential member either belongs or it doesn't, and none straddle the line.

For instance, suppose we want to find the set of long rivers. We go down the list of possible members—all rivers—and ask whether each is long. We get the following.

The Nile (4,180 m.) is a long river.	TRUE
The Hudson (306 m.) is a long river.	FALSE
The Danube (1,766 m.) is a long river.	TRUE
The Rhine (820 m.) is a long river.	FALSE
The Mississippi (2,348 m.) is a long river.	TRUE

And so on. The TRUEs—Nile, Danube, Mississippi, and other qualifiers—become the set of long rivers. A circle then seals them off.

The TRUEs and FALSEs together form the sphere of reference. Cantor called it the universe of discourse. As words need context to give them meaning, so do sets, and the universe of discourse provides it. Here, the universe of discourse is plainly all rivers. But often there is room for discretion, since context can vary. For instance, the set of ducks might be part of the universe of discourse of birds, or aquatic creatures, or perhaps all animals. It depends on what we are comparing the set to.

Cantor went on to move sets about like dancers at a minuet, probing the ways they could interact. Set theorists call these relations *operations*, and this stripped-down discussion addresses four of them: complement, containment, intersection, and union (see Figure 1).

■ *Complement: Who doesn't belong?* The complement of a set is its opposite. Thus,

| Set | Rivers that are long. |
| Complement | Rivers that are NOT long. |

In diagrams, this operation works like a cookie cutter on dough. Remove a circular set and the knot-holed leftover is the complement—the FALSEs, the rest of the universe of discourse.

■ *Containment: What groups belong to what other groups?* Like a Chinese box, a set can contain another set. It does so if it has every

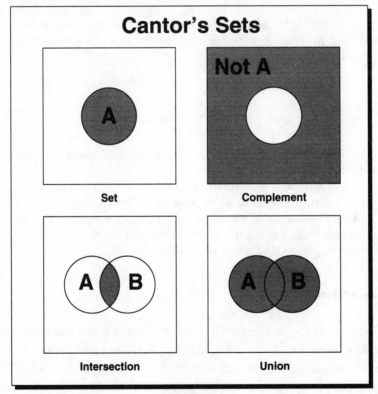

Cantor's Sets

Set

Complement

Not A

Intersection

Union

FIGURE 1

member of that other set. The smaller set is called the subset. For instance, the set of long rivers contains all the long Siberian rivers. *Long Siberian rivers* is thus a subset of *long rivers*, which in turn is a subset of *rivers*. Sets and subsets can create hierarchies of inclusion, pockets within pockets within pockets.

■ *Intersection: Who belongs to both groups?* When two roads cross, they create a square: the intersection. The intersection is part of both roads. Sets can cross in the same way. For instance,

Set	All golfers.
Set	All swimmers.
Intersection	All golfers AND swimmers.

Intersection weeds out and excludes. It accepts only the members of both sets. Thus Lee Trevino is in the intersection only if he is both a golfer and a swimmer. Mapped out in circles, intersection is the area where sets overlap.

■ *Union: Who belongs to either group?* Union merges sets without the filter. It is the big open hopper, swallowing every item that falls into either set. Thus,

Set	All golfers.
Set	All swimmers.
Union	All golfers OR swimmers (or both).

Lee Trevino is in the union since he golfs, and it doesn't matter whether he swims or not.

Trevino is also a set himself. A set can have one member. Thus, every set is the union of all its members. The set of tigers is the union of thousands of other sets, each a single creature. The set of shopping centers is the union of all shopping centers, and the set of spiral galaxies, the union of all spiral galaxies.

HAZE AT THE EDGES

Cantor's set theory went on to triumph among mathematicians, but a pesky paradox has always afflicted it, one that went back to the ancient Greeks. It was so well-known it earned a name: *sorites* ("suh-REE-tees"), the paradox of the heap. It's simple. Take a grain of sand from a heap and you still have a heap. Take another grain from it, and it remains a heap, and so on. Eventually one grain is left. Is it still a heap? Remove it and you have nothing. Is that a heap? If not, when did it cease being one?

Paradoxes of this kind abound. Among the more famous are:

- *Falakros.* Pluck a hair from a normal man's head and he does not suddenly become bald. Pull out another, and a third, and a fourth, and he still isn't bald. Keep plucking and eventually the wincing man will have no hairs at all on his head, yet he isn't bald.
- *The Paradox of the Millet Seeds.* Drop a millet seed on the ground and it makes no sound. But then how can dropping a bushel of millet seeds make a sound, since it contains only millet seeds? (After Zeno the Eleatic.)
- *Theseus' Ship.* When Theseus returned from slaying the Minotaur, says Plutarch, the Athenians preserved his ship, and as planks rotted, replaced them with new ones. When the first plank was replaced, everyone agreed it was still the same ship. Adding a second plank made no difference either. At some point, the Athenians may have replaced every plank in the ship. Was it a different ship? At what point did it become one?
- *Wang's Paradox.* If a number x is small, then $x + 1$ is also small. If $x + 1$ is small, then $x + 1 + 1$ is small too. Therefore, five trillion is a small number, and so is infinity. (After mathematician Hao Wang.)
- *Woodger's Paradox.* An animal can belong to only one taxonomic family. Therefore, at many points in evolution, a child must have belonged to a completely different family from its parents. But this feat is basically impossible genetically. (After biologist John Woodger.)

In Cantor's theory, one resolves such dilemmas by fiat. One simply dictates a breakpoint. A certain number of grains constitutes a heap; that number minus one is not a heap. Of course, no human speaker ever uses *heap* so precisely, or could, and logicians like Charles Dodgson ("Lewis Carroll," 1832–1898) saw the difficulties this fiction raised. If a heap has vague boundaries, the assumptions of set theory drain away. Dodgson shrugged ruefully and said, Just draw a line somewhere and pretend. It seemed a tiny sacrifice for the convenience.

However, a few scholars sensed graver trouble here, and their work ultimately led to fuzzy theory. They began by examining vagueness itself, and found it not just a marginal annoyance, but a vast domain.

Vagueness is haze at the edges. Is this pile a *heap*? Is a painting on the wall *furniture*? Is Sugar Loaf a *mountain*? Such questions evoke the "Well . . . ," the hand tilting right and left, the reference to "a matter of definition." Technically, vagueness differs from ambiguity in that ambiguity involves two or more incompatible meanings. "I heard the charge" is ambiguous, since *charge* can refer to an accusation, a headlong rush, or a command. It is one of them, and more information can reveal which. With vagueness, however, more information does not help.

The first thinker to grapple seriously with vagueness was Charles Sanders Peirce (1839–1914), America's most innovative philosopher. Peirce's compass was remarkable. He advanced Boolean logic, and he—not Claude Shannon—first applied it to electric circuits: the secret of the digital computer. He launched pragmatism, introduced experimental psychology to the United States, probed the shape of the Milky Way, devised quincuncial map projection, prefigured mathematical economics, and contributed to numerous other disciplines. Yet he was unable to keep a regular university position, perhaps because of his unconventionality and wit. "His whole personality was so rich and mysterious that he seemed a being apart, a super-man," said his friend W. P. Montague. "I would rather have been like him than like anyone else I ever met."[10]

Peirce laughed at the "sheep & goat separators" who split the world into true and false.[11] Rather, he held that all that exists is continuous, and such continuums govern knowledge. For instance, size is a continuum, as sorites shows. Time is a continuum, so though an acorn eventually becomes an oak tree, no one can say exactly when. Speed and weight form spectrums, as do effort, distance, and intensities of all sorts. Politeness, anger, joy, and other feelings and behaviors come in continuums. Consciousness itself is a continuum, varying not only in a single person, from high alertness through coma, but also across species, from humans to protozoans.

Hence, Peirce asserted that vagueness is a ubiquitous presence and not a mark of a faulty thinking. Words do not suddenly cease pertaining at points on the spectrum, but rather shade away. This kind of uncertainty will always afflict us. "Vagueness," he noted, "is no more to be done away with in the world of logic than friction in mechanics."[12]

Peirce even attempted to reconcile vagueness with logic. "Logicians have been at fault in giving Vagueness the go-by," he observed,[13] and in 1905 stated, "I have worked out the logic of vagueness with something like completeness."[14] Scholars have searched his posthumous papers in vain for this construct, and a true logic of vagueness would await Lotfi Zadeh.

Bertrand Russell (1872–1970) also probed this topic. In 1923, he published a short paper in which he said that both vagueness and precision were features of language, not reality. "All language is vague," he claimed.[15] For instance, *red* and *old* were clearly hazy. But so were apparently exact terms like *meter*. A meter was the distance between two marks on a metal bar in Paris. However, he said, these marks weren't dimensionless points, but patches of some thickness, so the distance between them wasn't crisp. The bar also expanded and contracted with heat and cold. Thus, *meter* is vague, though much less so than *red*. *Second* is also blurred. A second is a fraction of one rotation of the earth, but the earth is not rigid, its parts don't all rotate at the same speed, and any measurement of it is prone to error.

What about *true* and *false*? Without precise symbols, they too are vague. Hence, every proposition has an ill-defined range of facts which will make it true. "This is a man" may apply plainly to Smith or Robinson, but uncertainly to an adolescent or a Neanderthal. "This is poetry" may apply fully to Keats's odes, only marginally to dada or computer verse.

He went further. "Vagueness, clearly, is a matter of degree," he said.[16] A small-scale map is vaguer than a large one. A blurred photo of Jill or Jane is vaguer than a sharp one. A person we see at 200 yards is vaguer than at two feet. A drop of water before our eyes is vaguer than one under a microscope.

Over the next 40 years, other thinkers recognized the ubiquity of vagueness. Albert Einstein said, "As far as the laws of mathematics refer to reality, they are not certain, and as far as they are certain, they do not refer to reality."[17] Quantum physicist Louis de Broglie expressed a similar sentiment, as did philosophers W. V. Quine and Ludwig Wittgenstein.

THE WEDGE AND THE SPECTRUM

But Jan Łukasiewicz ("yahn woo-ka-SHEH-vitch," 1878–1955) took the first step toward a formal model of vagueness. He was a Polish logician and philosopher who grew up in Lvov, taught at the University of Warsaw, and served as minister of education in the 1919 cabinet of

Ignacy Paderewski, the pianist. In 1939, invading Nazis razed his home and destroyed his papers, and he took work in the city archives and lived in squalor. Foreseeing the bloody Warsaw Uprising of 1944, in which a quarter of a million people died, he and his wife fled to Münster, Germany, and dwelt in ruined cellars during the worst Allied bombing. He survived to accept a post at Dublin's Royal Academy of Science, where he taught till his death.[18]

Łukasiewicz invented the substructure of fuzzy sets, an early logic based on more values than true and false.[19] In 1920 he published a brief paper describing his new creation. In it, 1 stood for true and 0 for false. But, in addition, 1/2 stood for possible. A statement could have any one of these values. It seems like a simple step. But as soon as we take it, we unmoor ourselves and drift out into strange and interesting waters. As Łukasiewicz said, "Logic changes from its very foundations if we assume that in addition to truth and falsehood there is also some third logical value or several such values."[20]

For instance, it led to the apparent absurdity of opposites equaling themselves. Traditional propositional logic dealt with statements like

> It is true that snow will fall tomorrow,

whose negation is

> It is true that snow will not fall tomorrow.

Łukasiewicz added another kind of statement:

> It is possible that snow will fall tomorrow.

This statement has a value of 1/2. Its opposite is

> It is possible that snow will not fall tomorrow.

This statement also has a value of 1/2. Since of course 1/2 = 1/2, we have STATEMENT = not-STATEMENT. Assertion and opposite are equivalent.

Most people do not find it shocking. If a house is half-built, it is also half-unbuilt. If it is possible snow will fall tomorrow, it must be possible that it won't. These are partial contradictions, partly true and partly false. Such statements compel their opposites. If a cup is half-full, it must be half-empty.

Partial contradiction is not total contradiction. If a cup is half-full and half-empty, it is not both full and empty. One may wonder how anyone could make this mistake, yet history abounds with thinkers who have and gyrated in wild circles as a result, as we'll see.

In logic, the operation called *negation* defines opposites. It turns a statement into its antithesis. In true/false logic, the true (1) becomes false (0) and the false, true. A little table sums it up:

STATEMENT	NEGATION
1	0
0	1

In the three-valued logic of Łukasiewicz, the table simply gains another line:

STATEMENT	NEGATION
1	0
1/2	1/2
0	1

The values for binary logic remain intact at the corners. Hence, Łukasiewicz observed that his logic always held good for classical logic as well. It had to, since it contained 0 and 1. Three-valued logic does not unthrone two-valued logic, but broadens it.

The third value acted as a wedge, cracking the true/false vise apart. Once the breach opened, Łukasiewicz saw no reason to insert just one extra value. He could add four, seven, fifteen, as many as he liked. In fact, he could have infinite values, all strung out between 0 and 1, each a degree. True and false would bookend the vast inner range. Such a spectrum, he felt, plainly surpassed three-valued logic, or any other kind.

It was a critical insight. The rules of two- or three-valued logic might be misleading, because they addressed only a fragment of the domain, but laws that held for all values were, ipso facto, always good. Whatever their other drawbacks, they could not be too narrow.

This sliding scale also yielded greater precision. Instead of merely acknowledging an jntermediate value, multivalued logic conveyed its size. It could therefore quantify degrees of truth. For instance, rather than just indicating whether a river is long, it could show how long it was. With 1 standing for completely true and 0 for completely false, this approach yielded tables like the following:

The Nile (4,180 m.) is a long river.	1
The Hudson (306 m.) is a long river.	0.2
The Danube (1,766 m.) is a long river.	0.7
The Rhine (820 m.) is a long river.	0.4
The Mississippi (2,348 m.) is a long river.	0.8

Such values demanded a new kind of negation, and it embraced and formalized partial contradiction. Here, the opposite of a part truth is the additional amount needed to make an entire truth. Thus, a sampling of the truth table—the whole is now endless—might include

STATEMENT	NEGATION
0.05	0.95
0.30	0.70
0.63	0.37
0.91	0.09

If "The Rhine is a long river" is 0.4 true, then "The Rhine is not a long river" is 0.6 true. Numerically, the statement and its opposite add up to 1.

Once the lid was off, multivalued logics proliferated. The one-armed Emil L. Post (1897–1954) outlined a three-valued logic in 1921, a year after Łukasiewicz. Soon Kurt Gödel, John von Neumann, and Donald Kleene offered their own multivalued logics. The Russian Dmitri Bochvar built one that treated meaningless statements, and Hans Reichenbach and Zygmunt Zawirski pioneered another based on probability. Like non-Euclidean geometries, these logics could come in as many forms as the deviser could think of, as long as they were consistent. They made logic a kaleidoscope.

IN THE MUSEUM OF APPLIED LOGIC

Max Black (1909–1989) was the Leif Erikson of fuzzy sets, the man who first saw the shoreline and pointed the way. Coincidentally born like Zadeh in Baku, Black emigrated to England, earned his Ph.D., and took a faculty post at Cornell, where he taught for the rest of his life. A photo in later years shows a man with closely cropped hair and black-rimmed glasses, sucking meditatively on a pipe whose bowl he grasps with his hand. He is frowning slightly, as if listening to an argument he disagrees with, and it is an appropriate expression for a man whose final book was *The Prevalence of Humbug*.

Black outlined his proto-fuzzy sets in a 1937 article.[21] He agreed with Peirce that vagueness stems from a continuum, and with Russell that it has degrees. In fact, he said, a continuum implies degrees. The continuum need not actually be continuous. It can be discrete, like a dotted line, and if the intervals are small enough, they will escape notice. For instance, dialects of German change slightly from village to village across the countryside. The difference is negligible from one town to the next, but it slowly accumulates until speakers at opposite ends of the country cannot understand each other.

Or imagine, he said, an exhibit in a hypothetical Museum of Applied Logic. A line of countless "chairs" stretches toward the horizon. At one end is a Chippendale. Next to it is a near-Chippendale, virtually indistinguishable from the first. Succeeding items are less and less chairlike, until the row finally ends in a lump of wood. A normal person not only cannot draw a clean line between *chair* and *not-chair*, but performs the task uneasily. The concept *chair* does not permit this distinction.

Black observed that if a continuum is discrete, like the line of chairs, one can pin a number to each element. The number will indicate a degree, and a word like *chair* is the collection of these degrees.

But degrees of what? At this crux, Black mis-stepped. He could have chosen degrees of truth, but instead he opted for usage. The number on each of his objects showed the percentage of people who would call it a *chair*. A chair is a chair to the extent society calls it one. Vagueness was thus a matter of probability. If 67 percent of the population deems an item a chair, it is a chair to 0.67 extent. And the chance of the next person calling it a *chair* is also 67 percent.

In an appendix, almost as an afterthought, Black suggested that vague terms could form sets and display the panoply of set operations. For instance, they could have subsets. The populace always uses a subset word less often than a set word to describe the same item. Fewer people will call any object a *Chippendale* than a *chair*, or *enormous* than *big*, so the former is a subset of the latter. He also observed that one set could be a partial subset of another. Then he dropped the matter. He had glimpsed a new intellectual structure, but he never fit the pieces together.

THE FUZZY SYNTHESIS

If Max Black is the Leif Erikson of fuzziness, Lotfi Zadeh is its Christopher Columbus. Yet even this comparison falters, for Zadeh not only rediscovered this territory, but identified it, explored it, mapped it, promoted it, and fought for it. He is part Columbus, but also part Coronado, Franklin, and Washington.

In his 1965 paper, Zadeh assembled the parts of the puzzle and explained its importance.[22] Unlike Łukasiewicz and Black, he correctly addressed vagueness and set forth the mechanics of fuzzy set theory. It fuses classic sets and the logic of Łukasiewicz, and in fact resembles the New Math with decimal fractions.

The key notion, Zadeh's sudden insight in 1964, was graded memberships. A set could have members who belonged to it partly, in degrees. The classic example in the fuzzy canon is *tall men*. Suppose Tom is 5'11". Does he belong to *tall men*? Conventional sets ask, "Is Tom tall?" and erect a fence at, say, 5'10". *Tall* lies above this height, *not tall* below. Thus, at 5'11", Tom is tall. In contrast, fuzzy sets ask, "How tall is Tom?" The answer is a partial membership in the fuzzy set, such as 0.6. So Tom is 0.6 tall.

Thus, candidates for *tall men* might have the following memberships:

	TRUTH VALUES	
	CRISP	FUZZY
Jim (6'6") is tall.	1	0.95
Jon (6'2") is tall.	1	0.8
Tom (5'11") is tall.	1	0.6
Bob (5'9") is tall.	0	0.4
Bill (5'6") is tall.	0	0.2

The crisp set contains Jim, Jon, and Tom. Though Bob is almost the same height as Tom, he is not in the set. The fuzzy set contains all five and shows the similarity between Tom and Bob. Fuzzy sets discriminate much better and supply more information. They are, ironically, more precise.

Likewise, the set of long rivers can be crisp or fuzzy:

	TRUTH VALUES	
	CRISP	FUZZY
The Nile (4,180 m.) is a long river.	1	1
The Mississippi (2,348 m.) is a long river.	1	0.8
The Danube (1,766 m.) is a long river.	1	0.7
The Rhine (820 m.) is a long river.	0	0.4
The Hudson (306 m.) is a long river.	0	0.2

With this structure, the questions that vex classic sets begin to fall away. If Jill plays tennis twice a year, is she a tennis player? If Marie lives half the year in France and half in Brazil, is she an inhabitant of France? Cantor's theory shrugs and asks you, like Pope Alexander VI, to split these worlds with a line. Fuzzy sets describe such cases as they are.

Fuzzy sets easily resolve the paradox of the heap. With each grain of sand removed, the heap has less membership in the set of heaps. It drops from 1.0 through 0.8 and 0.2 to, finally, 0. Fuzzy sets glide smoothly across the truth continuum.

But is the Rhine really a *long river* to only 0.4 degree? It depends on judgment and context. One must estimate the memberships in fuzzy sets, a subjective task. This fact may seem disturbing, but there is no way out. *Long river* is inherently subjective. If it were a crisp set, the job of allotting some rivers to *long river* and not others would also be subjective. Fuzzy sets, with their decimal values, yield better estimates than just 1 and 0.

Moreover, Zadeh said, people have a "remarkable ability" to assign these grades of membership. For instance, students can easily give professors a membership in the set of good teachers. "Indeed, the assignment would usually be arrived at almost instantaneously without any conscious analysis" of the key factors.[23] Scientists who disliked such reliance on intuition, he said, could also base the values on harder facts such as polls and consensus.

The master key to fuzzy sets is membership values. In fact, fuzzy sets are easier to grasp as sets of degrees rather than sets of members. The fuzzy set of *long rivers* is less a set of rivers than of river memberships. It is not Nile, Mississippi, Danube . . . , but 1, 0.8, 0.7. . . .

Fuzzy sets include crisp sets. A crisp set is just a fuzzy one with membership values of 1 and 0. Crisp sets suggest that the nub is the existence of membership, while fuzzy sets show it is the extent. If an item is in a crisp set, it must have a value of 1; if it's "in" a fuzzy set, it can have any value except 0. Hence, fuzzy operations always work with crisp sets. They just look different.

For instance:

■ *Emptiness: Which sets have no members?* A fuzzy set is empty if all candidates have zero membership. For instance, the set of Norwegians who have visited Mercury is an empty set. The set of every integer that is both above and below 100 is an empty set. The set of living griffins, of oceans beginning with X, of maple trees atop the Matterhorn—all are empty sets.

Empty fuzzy sets are scarcer than empty crisp sets, because fuzzy sets cast the net so much farther. For instance, the set of Shakespeare scholars on Tristan da Cunha may be an empty crisp set. But if a bright person on the island has read most of the plays, that individual might have a 0.1 membership in the fuzzy set and save it from vacancy.

■ *Complement: How much do items not belong?* In English, a *complement* fills up or completes. The complement of a glass of

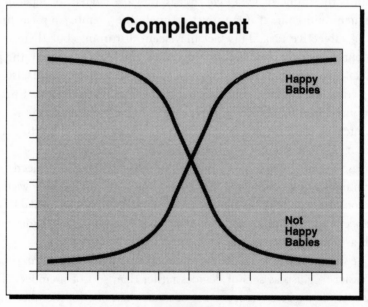

FIGURE 2

water two-thirds full is the one-third it needs to reach the top. Complements flesh out fuzzy sets the same way.

The complement of a fuzzy set is the amount the memberships need to reach 1. Suppose two people in a living room are watching *Bonfire of the Vanities* on the VCR. The set of annoyed people in the room is Sam/0.85 and Pam/0.80. Its complement, the set of not-annoyed people, is Sam/0.15 and Pam/0.20. That is, the complement is the extent to which each person is not in the set.

Max Black had depicted word usage as curves on a graph, and fuzzy sets form similar curves. For instance, Figure 2 shows the set of happy babies. The complement is the reverse of it, the set of degrees to which the same babies are not happy.

■ *Containment: What groups belong to other groups?* In crisp sets, a subset is fairly simple: Its members all belong to a larger set. All parakeets are birds. All novelists are writers. All neurons are cells.

Fuzzy sets, however, follow Black's standard: Each item must belong less to the subset than to the larger set (or equally to both). That is, fuzzy subsets have smaller memberships than sets, across the board.

Take the sets of *tall men* and *very tall men*. They might yield fuzzy membership values like the following:

	TED	JED	RED		ROD	TOD	TOM
Tall Men	1	1	0.9	...	0.7	0.6	0.55
Very Tall Men	1	0.9	0.8	...	0.5	0.4	0.3

Each membership in very tall men is equal to or less than its corresponding membership in *tall men*. So *very tall men* is a subset of *tall men* (see Figure 3).

Likewise, the set of *numbers close to 100* contains the subset of *numbers very close to 100*. The set of *hardy explorers* contains the subset of *extremely hardy explorers*. The set of *likely nominees* contains the subset of *highly likely nominees*.

■ *Intersection: How much are items in both sets?* In crisp sets, an intersection is the members two sets share. In fuzzy sets, it is the degrees of membership two sets share. A fuzzy intersection is thus the lower of each item's memberships in both sets. For instance, the intersection of *tall men* and *fat men* might look like the following:

	TED	JED	RED		ROD	TOD	TOM
Tall Men	1	1	0.9	...	0.7	0.6	0.55
Fat Men	0.2	0.5	0.1	...	0.8	0.3	0.9
Intersection	0.2	0.5	0.1	...	0.7	0.3	0.55

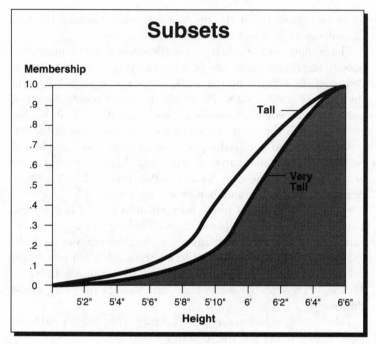

Subsets

Membership

FIGURE 3

The rule satisfies crisp sets too. If Hannah is a golfer (1) but not a swimmer (0), we take the minimum of the two—0—and give her zero membership in the intersection. But the fuzzy intersection reveals more. If she golfs every weekend (0.9) but swims only occasionally (0.2), she is in the intersection to 0.2 degree.

Fuzzy intersections are the realm of apparent paradox. For instance, they easily yield the set of men who are both tall and short—the shaded spire in Figure 4. Of course it is really an intersection of men who are partly tall and partly short. The contradiction arises from assuming crispness.

■ *Union: How much are items in either set?* The union of fuzzy sets is the reverse of the intersection. Instead of the lower of the two fuzzy values, it is the higher. The union is thus the larger of each individual's membership in either set. For instance, the union of *tall men* and *fat men* is the following:

	TED	JED	RED		ROD	TOD	TOM
Tall Men	1	1	0.9	...	0.7	0.6	0.55
Fat Men	0.2	0.5	0.1	...	0.8	0.3	0.9
Union	1	1	0.9	...	0.8	0.6	0.9

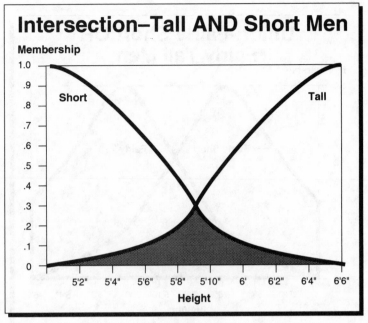

Intersection–Tall AND Short Men

Membership — Short — Tall — Height

FIGURE 4

Or, as Figure 5 shows, the union of *fairly short men* and *fairly tall men* includes everything beneath the gullylike profile.

Zadeh's initial paper made one more major contribution. It set forth the theoretical basis for fuzzy computer chips, which would not appear for another 20 years. A normal chip is made up of thousands of switches—transistors. Like the switches in a house, these either let electric current pass or stop it. They correspond to the true and false of Cantor's set theory. Zadeh proposed "sieves" instead. A sieve would let differing amounts of current pass, say, in quantities from 1 to 10. It would register more information than an on/off switch.

THE GRAYSCALE WORLD

Glance at a shadow and it will seem crisp. Look more closely and a zone of gray will appear at the fringe, a penumbra. Every shadow has one to some degree. Fuzziness is likewise ubiquitous, and the only reason people don't notice it is that they don't look for it.

For instance, Zadeh said, most words are fuzzy. They include any

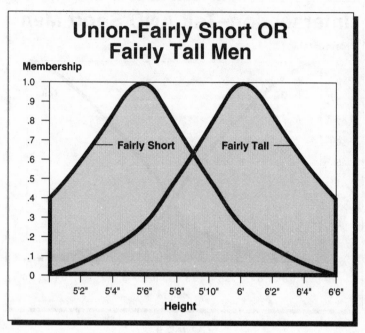

FIGURE 5

adjective or adverb. Some of these are obvious—*warm, deep, high, lithe, easily, gently*. But what about the absolute adjectives, like *full, flat, accurate, pure, equal, perfect*, and *absolute* itself? Such terms have both a crisp and a fuzzy sense, and the latter is by far the more common. For instance, tabletops have tiny irregularities and ornamental grooves, yet people routinely deem them flat. A city street is rougher than a tabletop and slightly arched to reject rainfall, yet we call it *flat* as well. The bottom of a foot, the broad top of a rock, the Middle West—all are even less flat, and still the term applies. Flatness, fullness, accuracy, and purity all admit of degrees, of comparisons. So do equality, perfection, and absoluteness. Phrases like *fairly equal* and *highly accurate* do not offend the ear, and the preamble to the U.S. Constitution speaks of a *more perfect union*.

What about *unique*? It means "being without a like or equal," and grammar texts state that, despite usage, one item cannot be *more unique* than another. But this dictum is arbitrary. Every person is technically unique, but some have features that set them apart more strikingly than others. The world heavyweight boxing champion, the president of the

United States, the richest person on earth—such individuals are unique to an extent that most people are not. Uniqueness too comes in degrees.

And *true*? Some logicians object to levels of truth. Wilfrid Hodges states, "*More true* can only mean 'more nearly true' or 'nearer the truth.' In this sense, there are no degrees of truth. Truth is absolute."[24] No matter how closely a proposition approaches truth, it is false until it crosses the line, when it becomes 100 percent true.

In nearing truth, however, statements partake of truth. It is rather like the border between the United States and Mexico. The political boundary is quite crisp, but for 50-odd miles on both sides, a hybrid culture exists in which residents speak English and Spanish. This zone differs in quality from either the United States or Mexico. One is not just near Mexico, but partly in it. Likewise, the statement "Sample X is a pure water" acquires more truth as the sample becomes purer and purer.

Of course, nouns and verbs can also be fuzzy. In his Museum of Applied Logic, Black showed how *chair* admits of degrees. A clumsy high school orchestra attacks *Für Elise*, and its *melody* remains, to some extent. And the merging overlap of *jog* and *run*, *interest* and *absorb*, *influence* and *cause*, shows that verbs too are a matter of degree.

Like Bertrand Russell and Carl Hempel, fuzzy theorist Bart Kosko says everything is fuzzy except numbers. For instance, take *alive*. Viruses can form crystals and cannot reproduce on their own. Are they alive? Or consider *this cabinet*. Molecules are constantly moving away from its surface for a moment, then rejoining it. Are they part of the cabinet? Kosko says they lie in the gray zone. The cabinet is fuzzy, though imperceptibly so.

Fuzziness seeps into everything and often appears in surprising places. As Kosko notes, "If fuzziness exists, the physical consequences are universal, and the sociological consequence is startling: Scientists, especially physicists, have overlooked an entire mode of reality."[25]

And not just scientists. Most citizens have long assumed inherent virtue in crisp categories. For instance, critics have ridiculed Los Angeles architects for fusing a welter of styles, not for fusing them gracelessly. Intriguingly, this trait may be ebbing. Some observers contend that we are more and more fuzzifying the old classes, dissolving the walls, grayscaling the world. Anthropologist Clifford Geertz speaks of new "blurred genres" in scholarship, fields like cognitive science, molecular biology, bioethics, and algebraic geometry,[26] as well as class-straddling works like Claude Lévi-Strauss's *Tristes Tropiques* and the scientific belles lettres of Stephen Jay Gould and Lewis Thomas. Writers of "faction" and gonzo journalism have turned the old fiction/nonfiction dichotomy into a gradient. Musicians are merging idioms, so that Philip Glass and Robert Wilson's *Einstein on the Beach*, for instance, combines classical, popular, and modern. Restaurants are blending French, Japanese, California, Chinese, and other cuisines. Even political units are growing fuzzier. The European Community is gradually reducing the degree to which France, say, is a *nation*. "We are seeing not just another redrawing of the cultural map—the moving of a few disputed borders, the marking of some more picturesque mountain lakes—but an alteration of the principles of mapping," says Geertz. "Something is happening to the way we think."[27]

PRECISION'S BACKUP

Complex systems had triggered the idea of fuzzy sets. How could fuzziness deal with them?

In the 1930s, philosopher A. Cornelius Benjamin had observed that simple concepts are more apt to be crisp and complex ones to be vague.[28] For instance, as Zadeh pointed out, the words on the left refer to more complicated notions than those on the right:

Oval	Straight line
In love	Married
Friend	Brother
Masculine	Male

The left column is fuzzy, the right one more or less crisp.

Zadeh realized that complex disciplines teem with fuzzy concepts. They include such notions as obscenity and insanity in law; arthritis, arteriosclerosis, and schizophrenia in medicine; recession, value, and utility in economics; grammaticality and meaning in linguistics; stability and adaptivity in system theory; truth, morality, and causality in philosophy; and intelligence and creativity in psychology. Fuzzy sets can describe them all.

The role of fuzziness in managing complexity lay implicit in the 1965 paper, but he developed it at length in the following years. The cornerstone was his Law of Incompatibility:

As complexity rises, precise statements lose meaning and meaningful statements lose precision.

Or, as he put it more formally: "As the complexity of a system increases, our ability to make precise yet significant statements about its behavior diminishes until a threshold is reached beyond which precision and significance (or relevance) become almost mutually exclusive characteristics."[29]

The Law of Incompatibility may seem to impose harsh conditions, because it places limits on precise analyses of complex systems. But it is also very practical. It provides a backup for precision when precision gets unwieldy.

When people face complex information, they trot out a handy strategy: summarization. Though a résumé lists pages of detail, the personnel manager may condense it into: "She's smart, but somewhat inexperienced." A commuter gazing out the window at cold, blustery rain thinks: "Bad weather." A reporter covering a subtle political maneuver begins: "The budget committee may have dealt a setback to the administration yesterday." Such simplifications catch the gist and, Zadeh observed, are always fuzzy.

The brain itself is constantly summarizing sense data, reducing massive detail to chunks of perception. For instance, the million light-sensitive cells of the human retina take in far too much information for the brain to decode. Instead, it has tactics for sifting and reducing this flow. It summarizes. Thus, Gestalt psychologists found that we see an almost-closed circle as a complete one. We don't process each point on the circle, since this approach would clog our minds. We take a shortcut, noting a few parts of the figure and perceiving a circle. Though we may err occasionally, we gain overall from the trade-off in speed and economy.

Anyone who has ever proofread a document understands the phenomenon. The task is vexing and unnatural. The mind normally races over sentences, picking up just enough information to extract the meaning and ignoring solecisms like *the the* and *conditon*. They don't quite reach the brain, and they shouldn't. They're excess.

Thus, we cut the flood of information down to the trickle we need. We round off the gapped circle and the *the the*. We perceive the precise in a fuzzy way. Zadeh called this ability one of the most important that we possess, and noted that it marks off living intelligence distinctly from that of machines.

It also helps explain the fuzziness of words. *Chair* distills an array of objects into one notion. *Furniture* summarizes even more broadly. Words centralize concepts that may have blurred bounds. Thus, Zadeh said language was "a system for assigning atomic and composite labels (i.e., words, phrases, and sentences)" to fuzzy sets.[30] Language is a vast shorthand, the outstanding instance of our ability to summarize.

Zadeh felt fuzzy logic could handle complexity in a similar way. As members in a class grow, they eventually exceed human comprehension. The brain responds by summarizing the class into chunks, labeled with words. For instance, it might divide the myriad hues of the spectrum into red, orange, yellow, green, blue, purple, violet, and other categories. Because each of these subclasses is a fuzzy set with degrees of membership, numbers can describe them. By summing up words mathematically, fuzzy sets could help bring complex systems like the visual apparatus under control.

He worked these ideas out in his key paper and, in the early fall of 1964, submitted it to *Information and Control*, on whose editorial board he sat. Twenty-seven years later, speaking in Tokyo in 1991, Zadeh said he had recently talked to Murray Eden, former editor of that journal. "He confirmed what I thought," Zadeh said, "that my paper was published only because I was a member of the editorial board."

It was an omen.

2

THE COCAINE OF SCIENCE

■

> What makes society turn is science, and the language of science is math, and the structure of math is logic, and the bedrock of logic is Aristotle, and that's what goes out with fuzzy.
>
> —BART KOSKO

> The study of the basic philosophies or ideologies of scientists is very difficult because they are rarely articulated. They largely consist of silent assumptions that are taken so completely for granted that they are never mentioned. . . . [But] anyone who attempts to question these "eternal truths" encounters formidable resistance.
>
> —ERNST MAYR

> I was fully cognizant that I was doing something that would spark controversy.
>
> —LOTFI ZADEH

■

Zadeh previewed his first paper to a gathering of some 20 Berkeley professors in November 1964. "Let's put it this way," he says. "Of those who heard my presentation, those who expressed themselves, they were negative." They were not insulting or combative—these were Zadeh's friends—but they were critical. Fuzzy logic was under fire and it had not even appeared in print.

Not everyone responded with a show of fangs. In 1967 Max Black wrote Zadeh, "Now that I have had a chance, at last, to study your work, I want to express my admiration and interest. I believe that your ingenious construction promises to provide intellectual tools of great value."[1] Zadeh also proved the center of warm interest at a 1968 Honolulu conference on pattern recognition.

However, he faced no dearth of criticism. For instance, at a November 1966 colloquium at UCLA, Arthur Geoffrion, a well-known scholar in decision analysis, asked rhetorically, "Where do the membership values come from?" Out of our heads, he said. They were subjective. Fuzzy theory offered no way to pinpoint them objectively, he said, yet it required this ability.

At a 1972 conference in Bordeaux, Zadeh clashed with Rudolph E. Kalman, inventor of the Kalman filter, a major statistical tool in electrical engineering. Among other feats, the Kalman filter enables accurate targeting of moving objects, and Patriot missiles in the Gulf War used it to shoot down Iraqi SCUDs. A former student of Zadeh's, Kalman objected to fuzzy logic on the following grounds:[2]

First, he said, the great goal of system theory was a skeleton key to the workings of systems, analogous to Newton's laws. Fuzziness was hardly this grail. For instance, Kalman observed, modern research has revealed "that the brain, far from fuzzy, has in many areas a highly specific structure." Neuroanatomists were making rapid progress with the electron microscope, a new tool that was "clarifying regularities of structure which previously were seen only in a fuzzy way."

Kalman also said, "No doubt Professor Zadeh's enthusiasm for fuzziness has been reinforced by the prevailing political climate in the U.S.—one of unprecedented permissiveness. 'Fuzzification' is a kind of scientific permissiveness; it tends to result in socially appealing slogans unaccompanied by the discipline of hard scientific work and patient observation. I must confess that I cannot conceive of 'fuzzification' as a viable alternative for the scientific method." Zadeh treasures this quote so much he has made a slide of it for speeches.

Fuzzy logic was also off-beam pragmatically, Kalman said, since there was no evidence it could solve important problems. Yet engineers had gained excellent results by using precise means like probability to deal with uncertainty. Kalman did not claim fuzzy tools would never be useful. "But if one proposes to deprecate [precision] . . . he should at least provide some hard evidence of what can be gained thereby," he said. "Professor Zadeh's fears of unjust criticism can be mitigated by recalling that the alchemists were not prosecuted for their beliefs but because they failed to produce gold."

A few years later, in 1975, William Kahan, a mathematician and colleague of Zadeh's at Berkeley, added three more contentions:[3]

- "I can not think of any problem that could not be solved better by ordinary logic."
- "What Zadeh is saying is the same sort of thing as, 'Technology got us into this mess and now it can't get us out.'"
- "What we need is more logical thinking, not less. The danger of fuzzy theory is that it will encourage the sort of imprecise thinking that has brought us so much trouble."

Kahan would maintain these opinions almost intact over the years. "I'm not an expert in this area," he cautioned in 1989. "If you were to quote me, you're quoting someone on an area I decided 20 years ago I didn't want to work in."

He maintains that fuzzy logic grins at inconsistencies. "With traditional logic there is no guaranteed way to find that something is contradictory, but once it's found, you'll be obliged to do something. But with fuzzy sets, the existence of contradictory sets can't cause things to malfunction. Contradictory information doesn't lead to a clash. You just keep computing." Here, an intelligent person is confusing partial contradictions with polar ones.

Kahan believes fuzzy logic affords a treacherous, wobbly base for science. "Life affords many instances of getting the right answer for the wrong reasons," he says. "It is the nature of logic to confirm or deny."

Kahan will entertain insights obtained by invalid or illogical thought, but he goes back carefully to see if he can verify them with logic. "The fuzzy calculus blurs that."

Like Geoffrion, Kahan contemns the assignment of truth-values as too subjective. Probability gives assertions about randomness a "quantitative meaning that can be verified. The house at Vegas makes money, or the insurance company does. But there's no way to verify the fuzzy logic, because there's no way to assign numbers to the initial data. So you choose numbers because you think it'll make the system work. Scientific proof has to be falsifiable in principle or else it isn't true."

But doesn't fuzzy logic work? Like other critics, Kahan admits that it may succeed most of the time, but "there will inevitably arise exceptional situations and then what the system does will be hard to predict. And if you believe in fuzzy logic, you won't blame it. You'll have to blame something else."

Kahan concludes, "Logic isn't following the rules of Aristotle blindly. It takes the kind of pain known to the runner. He knows he is doing something. When you are thinking about something hard, you'll feel a similar sort of pain. Fuzzy logic is marvelous. It insulates you from pain. It's the cocaine of science."

Zadeh bore such barbs with remarkable ease. "In the early days he received all kinds of criticism in words and writing, some very harsh," says John Yen, a professor at Texas A&M. "So he says that throughout the years he learned to take whatever people said to him as a compliment."

"As a person, he's a model of what science is about," says logician Brian Gaines. "It's about what's in your head and the accurate, honest expression of that. Criticism is irrelevant except at an intellectual level."

Kosko adds, "Zadeh was chairman of the top electrical engineering department for some years and then he published fuzzy and he took a lot of shit. If anyone can handle it, he can."

Indeed, Zadeh professes utter equanimity to the negative reaction. "There were many, many other issues where I was arguing in favor of something that most people were against. So this was not a new experience.

"I think generally I stand up for what I believe in," he says. "That's one of my characteristics." At the same time, he understands the power of diplomacy and avoids confrontation unless the situation compels it. "In so far as my disagreements over technical issues, I've never taken these things personally. So that many of the people who have been and still are very critical of fuzzy sets are good friends of mine."

Despite this geniality, fuzzy logic could kindle a fiery scorn, a reaction seemingly out of proportion to the idea. What caused it?

WHAT'S IN A NAME?

The first provocation was the name. Part of Kalman and Kahan's hostility seems to spring from the tag *fuzzy logic* itself, which both misdirects the attention and seems to celebrate mental fog. They were not alone.

Why *fuzzy*? Zadeh says the term was concrete, immediate, and descriptive. "I realized, of course, that many people would be repelled by the word *fuzzy*, because it's usually used in an uncomplimentary sense, a negative sense." In fact, it had exactly this effect on many. "I knew that just by choosing the label *fuzzy* I was going to find myself in the midst of a controversy."

Why *logic*? Fuzziness rests on fuzzy sets, and fuzzy logic as logic is really a small part of the field. Why use it? "Somehow to laymen, it's difficult to figure out what fuzzy sets are," he says. "*Fuzzy logic* is a little more self-explanatory, so I've been using the term *fuzzy logic* in a broader sense."

He would not have done it differently. "If it weren't called *fuzzy logic*, there probably wouldn't be articles on it on the front page of the *New York Times*," he says. "So let us say it has a certain publicity value. Of course, many people don't like that publicity value, and when they see it in the *New York Times*, it doesn't sit well with them." He feels that the negative overtones of *fuzzy* will wane with usage, and that, overall, the good has outweighed the harm.

Has it? No one can calculate how many scientists and corporations have ignored fuzzy logic because of the name alone. "My company reacted like nearly everyone in the United States and Europe," says Andrew Whiter, an engineer at Hewlett-Packard, one of America's premier electronics firms. "The term puts people off from the outset. They can't believe it's to be taken seriously."

"People hate to be associated with something not very precise," says Frank Wu, of the corporate research and development (R&D) lab at Motorola, America's largest independent maker of computer chips. "Fuzzy tends to be regarded as kind of sloppy, like you're doing a half-assed job, even though it's not true. Professor Zadeh may have picked the wrong word."

"I think people never took it seriously," says Ron Yager. "It's a name that lends itself to humor. I remember there was a guy with an office next to mine, and he was always saying, 'How are the fuzzy-wuzzies today?' "

John Dockery, who worked for the army chief of staff in the Pentagon in the early 1970s, tried and failed to interest the brass in fuzzy logic. "These were well-meaning people," he says. "But they reflected the reaction of the great mass of engineers and decision makers that fuzziness is not a good thing. That it's a term for children, or bears. If a commander said he had a fuzzy thinker on his staff, his career would be over."

The name swapped credibility for a public spotlight. However, if some fault is Zadeh's, much of it clearly rests with the engineers and corporations themselves. Science is a matter of truth, not appearances. Why spurn an idea because of its name? Why even form an opinion about it?

Yet Zadeh would have met obstacles even if he had given it an apt, neutral name like "continuous set theory." As he observed, "The concept of a fuzzy set has an upsetting effect on the established order."[4] Indeed, it challenged one of the oldest, deepest intellectual biases in the West.

THE SHADOW OF ATHENS

A few years ago, one of the authors happened to be dining in a Beverly Hills restaurant near Mel Brooks. The waitress appeared and listed the evening specials for him. One appetizer, she said, was yellowtail grilled on one side and raw on the other. "Hey, what is this? It's either *sushi* or it *isn't!*" he cried, articulating a 2,300-year-old law of classes straight out of Aristotle.

Georg Cantor did not originate the notion of classes as silhouettes.

The Greeks did. They ordained that a class has sharp boundaries. A pile of sand is a heap or not a heap. A man is bald or not bald. A number is small or not small. This idea came to permeate Western thought as an assumption, the kind of intellectual floor one defends instinctively from fear of tumbling into the cellar. Indeed, it settled in so deeply that Hobbes, Descartes, Locke, Leibniz, and the early Wittgenstein accepted it as natural, and Georg Cantor took it for granted in creating set theory—despite the obvious problem of separating heaps from non-heaps. Said two American mathematicians in the 1950s, "No one will doubt that . . . the statement 'Either it is raining or not raining' is *necessarily true.*"[5] But everyone has seen the fine mist that is neither rain nor not rain.

It began with Aristotle (384–322 B.C.), who dominated Western thought longer than any other secular figure in history. Universal expert, tireless author, tutor of Alexander and local tyrants, school administrator, and possibly a secret agent, he yet remains a ghostly figure to us. The legend-loving Diogenes Laertius, writing some 500 years after his death, describes him as bald, thin-legged, small-eyed, lisping, and fond of opulent garb. The great sage is also rather witty. Why are beautiful women so sought after? "That is a question fit for a blind man to ask." Did someone insult him the other day? "He can hit me too, if he likes, in my absence."[6] It's all about as reliable as P. T. Barnum profiling Shakespeare.

In just 12 years at his school the Lyceum, he apparently spun forth the books by which we know him today. They cover a vast tract: literature, biology, physics, ethics, law, politics, metaphysics, and much else besides. Including logic.

"Few persons care to study logic, because everybody conceives himself to be proficient in the art of reasoning already," C. S. Peirce wrote. "But I observe that this satisfaction is limited to one's own ratiocination, and does not extend to other men."[7]

Aristotle pursued logic assiduously; indeed, he may have invented it. He not only set forth the underpinnings and inference mechanisms that would dominate formal logic for over 2,300 years, but also delineated logical classes, the boardpieces of his logic.

He derived these classes and formal logic itself from mathematics, which had held great sway over the Greek mind ever since Pythagoras (c. 585–c. 500 B.C.). Aristotle wanted to extend the step-by-step method of math to reasoning in general. Math yields certainty, and he sought the same reward from thinking overall. To some extent, he found it. For instance, one can schematize inclusion: If A contains B, and B contains C, one must conclude that A contains C. By this route, math became the template for logic. In 1689, John Locke (1632–1704) declared that we should manage every logical argument as a mathematical proof. The logic of George Boole (1815–1864) modeled arithmetic, the great undecidability proof of Kurt Gödel (1906–1978) mapped logic back onto math, and scholars still debate the primacy of the two.

The mathematical origin of logic had at least two upshots. First, math has an edificelike structure. It proceeds from the bottom up, deductively. On the base of a few rules and axioms, we can build a tower of certainty, as solid as its premises. Logic would also follow this plan. Second, the units of math are crisp. A number is either 2 or it isn't. Hence logical units—classes—would be crisp as well.

"It is not everything that can be proved, otherwise the chain of proof would be endless," said Aristotle. "You must begin somewhere, and you begin with things admitted but undemonstrable."[8] Thus Aristotle began his search for the bedrock of logic: truths so obvious we accept them without proof. These were the axioms.

The first was the Law of Contradiction. Succinctly, it is: "A cannot be both B and not-B." In the *Metaphysics*, Aristotle states it more carefully: "The same thing cannot at the same time both belong and not belong to the same object and in the same respect."[9] Thus, a sheep cannot be both white and not-white. At tomorrow noon Singapore cannot be hot and not-hot. A dish can't be sushi and not-sushi. "This," he adds, "is the most certain of all principles."

The second was the Law of the Excluded Middle—more properly, the Law of Bivalence. It states, "A must be either B or not-B." In the *Metaphysics*, Aristotle expresses it thus: "Of any subject, one thing must be either asserted or denied."[10] In other words, a sheep is either white or not-white. At tomorrow noon Singapore will be hot or not-hot. A dish is either sushi or it isn't. When William Kahan says, "It is the nature of logic to confirm or deny," he is avowing this principle.

The two laws have a similar clang, but note the difference. The Law of Contradiction forbids true and not-true at once. Opposites cannot overlap. The Law of the Excluded Middle forbids anything other than true and not-true. Opposites must abut. Hence, not only can't a sheep be *both* white and nonwhite, but it must be *either* white or nonwhite.

Together, these laws create a widthless line between two opposites. *Not-hot* embraces all the ground *hot* omits. This line marks off the silhouettes. If we ask, "Is Singapore hot?" the axioms compel a cutoff line between *hot* and *not-hot*—say, at 85°F. Thus Singapore is hot at 86°F but not hot at 84°F. Without such arbitrary distinctions, *hot* and *not-hot* blur together in soft focus, and both laws crumble.

Cantor's set theory obviously embodies these laws. For instance, a set and its cookie-cutter complement cannot overlap. Their intersection is the empty set, the set with no members. Thus,

Set	All jolly people.
Complement	All not-jolly people.
Intersection	Empty set.

This declaration reissues the Law of Contradiction in set theory terms. No one can be both jolly and not-jolly.

Likewise, the union of a crisp set and its complement is the entire universe of discourse. Hence,

Set	All jolly people.
Complement	All not-jolly people.
Union	All people.

This rule restates the Law of the Excluded Middle. Every person must be either jolly or not-jolly.

Fuzzy sets flout both laws. They let a person be jolly in degrees, so she can be partly jolly and not-jolly at once. Hence the intersection of a fuzzy set and its complement is not a void. Overlap abounds. Likewise, the fuzzy union of a set and its complement is the higher of each person's jolliness and not-jolliness, and it does not fill the universe of discourse.

Most scholars for the next 2,300 years accepted the axioms, partly on the word of Aristotle. Like Galen (c. 130–c. 200 A.D.) and Ptolemy (fl. 2nd century A.D.), he became for medieval Europe an authority unto himself, a huge barricade to new ideas. Aristotle had said large bodies fall faster than small ones. Lucretius and Leonardo believed it; Galileo disproved it and found the equation that describes the universal rate of fall. Aristotle said maggots in rotting meat arise from nowhere; Francesco Redi (1626–c. 1697) largely proved him wrong, but the theory of spontaneous generation persisted till the 19th century and Louis Pasteur. Aristotle divided animals with red blood from all others; Linnaeus (1707–1778) would provide the correct taxonomy. Aristotle claimed it made no sense to speak of bounds to infinity; Cantor would show it comes in many sizes.

However, Aristotle's imperium in logic survived long after it began toppling in the natural sciences. In fact, it always exceeded that in other fields, partly because he analyzed logic so minutely, but also because medieval Europe knew his logic and not his other work, and because formalisms like logic resist modification from facts. Under Aristotle's influence, Descartes, Locke, and Leibniz ("We are to hold to this above all: every proposition is either true or false"[11]) projected crispness onto philosophy, and in 1787 Immanuel Kant stated that, since Aristotle, logic "has not been able to take a single step ahead and, hence, to all appearance, seems to be finished and complete."[12]

But scholars also had practical reasons for accepting bivalence. By halving the universe, it greatly simplified logic. Two degrees—*true* or *false*—are far easier to handle than myriad degrees in between. The efficiency justified the fudging. Indeed, false precision is an age-old trick to smooth information handling. For instance, when Congress sliced the United States into time zones in 1883 to simplify train schedules, it made fuzzy values crisp—four clock times instead of a spectrum. Yet at the border, time leaps ahead an hour when you cross an imaginary line, and things feel unnatural. Aristotle created a similar effect in logic.

Bivalence had a second virtue, the proof called *reductio ad absurdum*. If we can take two statements and reason logically with them to a contradiction—A equals not-A—we can presume one is false. If we say "Bismarck was consul of the Roman Empire" and "Bismarck was chancellor of Prussia," we can reason to "Bismarck was born before 476 A.D." and "Bismarck was born after 476 A.D." Both cannot be right. Hence, at least one of the initial statements is wrong. This proof has been a boon, especially in math, though "intuitionist" mathematicians such as Luitzen E. J. Brouwer (1881–1966) have sought to limit its use.

Finally, bivalence fostered certainty. *True* and *false* are absolutes in Aristotle's system, with no taint of equivocation. And since the reward of logic was the certainty of its conclusions, given its premises, logicians wanted to start with sure premises. How can one reach certainty from a *partly true*?

ARISTOTLE VS. ARISTOTLE

But are the laws correct? For Aristotle, their power lay in their intuitive truth, their accord with our common sense. Yet Aristotle himself does not view statements as either true or false. In the *Metaphysics*, a few pages before announcing the Law of the Excluded Middle, he says, "The more and the less are still present in the nature of things" and adds that one who thinks 4 equals 5 is more correct than the one who thinks 4 equals 1,000.[13] In *De Interpretatione*, he uses the terms *truer* and *falser*. Even for him, truth has degrees.

In fact, Aristotle provides a wealth of arguments against his own axioms. The most famous involves ships clashing at sea. In Chapter 9 of *De Interpretatione*, he ponders the statement:

There will be a sea battle tomorrow.

True? Not true?

A determinist would argue that, at this instant, it must be either true or false. Events are preordained, by divine hand or physical law. If we knew enough about celestial will or the sequence of causes, we could tell right now whether it were true or false. We don't, but it doesn't matter. At this moment, the statement is either true or false.

A nondeterminist would demur. Events are not preordained, so at this moment it may be impossible for us to know the answer, even with total knowledge. God may reconsider or random acts may jostle causality. The answer may simply not exist yet. If so, the statement is currently neither true nor false, but indeterminate.

De Interpretatione is a dense, muddled work, and the crucial Chapter 9 is so opaque that scholars still contest its meaning. However, it appears to set forth these arguments, half-heartedly reject determinism, and conclude that "There will be a sea battle tomorrow" has an intermediate truth-value. A gap opens at the hairline. The Law of the Excluded Middle ebbs away.

This question exerted great historical influence. In the Middle Ages, both Christian and Muslim thinkers tended to accept only two values for the sea battle, yet some scholars took up where Aristotle left off. Duns Scotus (c. 1265–c. 1308) toyed with a third value straight from Aristotle, as did William of Ockham (c. 1300–1349), he of the famous razor. In the 20th century, this chapter would spur Jan Łukasiewicz to revolt against both laws.

The sea battle caused much spume to fly, but Aristotle made other important reservations too.

First, silhouette definitions are crucial to Aristotle's logic. Definitions *are* the hairline. They must cleanly separate items into *chair* and *not-chair*. Yet at the outset of *De Anima*, he admits the difficulty of finding any method of definition. It is a remarkable concession, for if we can't specify the hairline, classes blur at the edges and the whole two-valued system becomes an artifice. He glozes over this problem in his works on logic, most likely for pragmatic reasons.

In the *Nicomachean Ethics*, Aristotle simply jettisons the axioms when discussing his famous golden mean. Here, an array of qualities possess a middle, usually the apex of virtue. Take courage. It is a continuum. The craven, who fear too much, lie at one end and the foolhardy, who fear too little, at the other. A sensible person seeks the ground in between: valor.

But what are *cravenness*, *valor*, and *foolhardiness*? They resist iron encirclement, he admits. Yet he is not now in the realm of logic, but elsewhere, he says, and must speak in terms which thwart sharp definitions. In fact, they aren't even necessary. "Our statement of the case will be adequate," he declares, "if it be made with all such clearness as the subject-matter admits; for it would be as wrong to expect the same degree of accuracy in all reasonings as in all manufactures." With topics such as character and ethics, "we must be content to indicate the truth roughly and in outline."[14] There is no hairline. Concepts wash over into each other, *valor* subtly merges into *foolhardiness* (i.e., *nonvalor*), and both laws dissolve.

He goes even further. Excess specificity, he says, leads to error, for "an educated person will expect accuracy in each subject only so far as the nature of the topic allows."[15] In other words, some subjects are simply vaguer than others, and we must fit the level of precision to the topic at hand. This statement comes very close to Zadeh's Law of Compatibility.

APPLES IN THE REFRIGERATOR

Of course, on a day-to-day level, everyone constantly deals with propositions that are neither true nor false. Will it rain tomorrow? Regardless of determinism, the answer is uncertain for all practical purposes. Yet most people want to hear it, since the chance of rain affects their plans. In the latter 1600s, scholars began mathematicizing such uncertainties, and here too they followed Aristotle's laws. They created probability.

Probability is Aristotelian because it works with atoms of yes and no. A flipped coin is either heads or tails. A playing card falls into one of 52 categories and its identity is never vague. Thus, probability obeys the twin laws. A flipped coin can't be heads and tails, and it must be either heads or tails. A card can't be both the king of diamonds and not the king, and must be one of them.

In mapping out vagueness, Max Black employed probability, perhaps because it was the dominant uncertainty tool at hand. He stated that if 80 percent of people say Tom is *tall*, then he is tall to 0.8 extent. At bottom lies the yes/no atom: Is Tom *tall*?

This approach differs fundamentally from fuzziness and yields different results. Suppose Sally considers Tom 0.8 tall. She will probably always call him *tall*. Suppose a thousand people believe Tom is 0.8 tall. Ask them, "How likely are you to call Tom *tall*?" and every one might say, "100 percent." Yet they would all think he is 0.8 tall. Each is rounding off from 0.8 to 1, so the final figures describe word usage, not degree of truth.

Zadeh distinguished fuzziness from probability in his 1965 paper. Both describe uncertainty numerically, he said. However, probability treats yes/no occurrences, requires ignorance, and is inherently statistical. Fuzziness deals with degrees, does not require ignorance, and, he added, is completely nonstatistical.

For instance, to take Kosko's example, if we ask, "Is there an apple in the refrigerator?" we are dealing with probability. The answer might be 0.5, as in a coin toss. But suppose we know there is a half-eaten apple in the refrigerator. The question now becomes, "To what degree is there an apple in the refrigerator?" That is a fuzzy question. The answer, however, is still 0.5.

Unlike fuzziness, Kosko notes, "Probability dissipates with increasing information." The more we know about the inside of the refrigerator, the less uncertain it is. If we open the door and gaze within, it becomes certainty: yes or no. Probability vanishes. It simply requires ignorance. That is why so much of it involves the future. Fuzziness, however, can coexist with total information. We can know everything possible about the half-eaten apple in the refrigerator, and the fuzziness remains.

Fuzziness plainly differs from probability, but also plainly resembles it, since both deal with degrees—one of truth, the other of likelihood or expectation. Hence, given the momentum probability has built up over the centuries, there is now tumult and shouting over their relation. Probability is today the great rival of fuzzy logic, and its champions claim it surpasses fuzzy logic in any task one can devise.

THE PERSISTENCE OF PARADIGMS

If fuzzy sets were true and useful, one should expect scientists to welcome them. However, when an idea lies far enough outside the frame of reference, the opposite typically occurs. In general, scientists resist notions that challenge basic assumptions.

In 1962 Thomas Kuhn surprised the academic world with *The Structure of Scientific Revolutions,* which undermined the classic view of scientific progress so seriously that it has never recovered.[16] Previously, most people saw science as a rational accumulation of truths. Investigators conceived and tested theories which, if proved, entered the pantheon of facts. The process was fair and impartial, driven only by the urge to understand. By this model, fuzzy logic would come before an impartial tribunal and receive a wise and expert judgment.

Kuhn unveiled a new model: science as hermit crab, periodically discarding a shell and adopting a larger one. Facts might be facts, but the great vehicle of knowledge is the paradigm, the master interpretation of a field. The paradigm unites the facts in a single architecture. Imagine a field of dots, apparently random and devoid of pattern. If more dots began appearing, like newly discovered facts, one might soon be able to argue that they formed two heads in profile, gazing at each other. This paradigm could show researchers where to search for more dots—to the left of one profile and the right of the other. Indeed, it could predict these extremely well, and this success might allow it to prevail for a long time, even if a few dots appeared between the heads. But if enough such interfacial anomalies showed up, the two-head paradigm would weaken and the field enter a crisis. A new theorist might then propose that the illustration really showed a vase and that the dots tracked the shadows of its ribs. This interpretation is drastically different and many observers would spurn it, but if pro-vase evidence continued to mount, the scientific community would eventually accept it. Kuhn called such a changeover a *paradigm shift*.

Paradigm shifts have occurred fairly often in science. For instance, Ptolemy thought the earth was the center of the universe and proposed a celestial model in which sun, stars, and planets whirled around the earth once a day. This schema suggested questions, guided researchers, and increased knowledge. Yet as astronomers tracked the planets ever more closely, they found oddities of motion that the theory wheezed to explain. Trapped within it, they added more and more excrescences, like the multiplying branches of a bad alibi, until the sheer weight of them triggered a crisis. Eventually, Copernicus stepped forth with a neater paradigm: the sun-centered universe.

Replacing a paradigm, however, can be a long and bitter business. Of course, global ideas deserve much keener scrutiny than local ones, because of their consequences. Stars fly around the earth or they don't. Yet scientists tend to balk at new paradigms even if sound. Understanding presupposes a frame of reference, and often the old paradigm is the frame of reference. The incumbent judges the challenger. Moreover, scientists can have vast confidence in the old paradigm, an assurance on which they have staked careers and reputations, and which in fact has bred useful work. They see it as a cozy, even noble vision, not a dying illusion.

Examples of resistance abound. Copernicus, Newton, Lavoisier, Dalton, Darwin, Einstein—all met detractors and often acid controversy. When Cantor published his proofs about infinite numbers, some mathematicians called him a fraud, and Henri Poincaré declared, "Later generations will regard [his book] *Mengenlehre* as a disease from which no one has recovered."[17] The eminent Leopold Kronecker especially belittled him, and helped keep him at the lowly University of Halle all his life. Cantor eventually suffered several nervous breakdowns and died in a local asylum in 1918.

"Mathematicians, let it be known, are often no less illogical, no less closed-minded, and no less predatory than most men," says historian of math Morris Kline. "Like other closed minds they shield their obtuseness behind the curtain of established ways of thinking while they hurl charges of madness against the men who would tear apart the fabric."[18]

Likewise, scholars long treated multivalued logics as mere curios. Willard Van Normand Quine is one of the leading American philosophers of the late 20th century, and he understands the problem of vagueness. Nonetheless, in 1970, he termed multivalued logics "deviant," lamented their lack of simplicity and familiarity, and said, with regard to their repudiation of Aristotle's axioms, "It is hard to face up to the rejection of anything so basic."[19] Why would anyone support these logics? He had heard several reasons, he said, but the "worst one is that things are not just black and white; there are gradations. It is hard to believe that this would be seen as counting against classical negation; but irresponsible literature to this effect can be cited."[20] By 1987, however, he had retreated substantially, asking that we tolerate a "double standard,"[21] that we admit the ubiquity of vagueness but continue to accept bivalence on the basis of its elegance and simplicity.

Thus, by Kuhn's model, fuzzy logic would come before a tribunal prejudiced against it by the 2,300-year-old Aristotelian dichotomy and by its offshoot, probability. It could expect dismissals, derision, and an apparently mulish inability to understand.*

The Kuhn construct has proved sensational. His book has sold almost a million copies in 16 languages, and the concept of paradigm shifts has diffused throughout the culture, often mutating in ways that displease Kuhn. Some scientists still oppose it and other philosophers have modified it, but there seems little question that a politics of science exists and affects the course of an idea's acceptance.

Though a misleading name and entrenched paradigms tilted opinion against fuzzy logic, so did a further force, one involving the nature of human thought itself.

---

* Indeed, as a bitter Max Planck said, "A new scientific truth does not triumph by convincing its opponents and making them see light, but rather because its opponents eventually die, and a new generation grows up that is familiar with it."[22]

THE PANDA AND THE BARBER

In 1869, the French missionary Armand David first described the giant panda to the West. It was a kind of bear, he thought, and gave it the scientific name *Ursus melanoleucus*, or "black-white bear."[23] He sent its bones to Alphonse Milne-Edwards, later director of the Paris Natural History Museum, who deemed it more a raccoon. He called it *Ailuropoda melanoleuca*.

Which was it? Scholars debated the matter for the next 120 years, issuing over 40 theses with no clear consensus. By 1987, Stephen O'Brien and his colleagues analyzed its DNA and determined that it is more a bear, and that it diverged from the ursid evolutionary track between 15 and 25 million years ago, after raccoons had. The giant panda is actually not much of a bear or a raccoon. That, for bivalent humanity, was the problem.

Why have intelligent people ignored degrees of truth for so long? Bart Kosko suggests one reason is that we round off constantly, automatically, unconsciously. The panda did not round off easily, so the process reached the level of awareness and, indeed, dispute. Most of the time, however, it remains concealed. We pour a glass of water and call it *full*, though water doesn't reach the brim. (We round 0.9 to 1.0.) Then we drink it and call it *empty*, though a few drops remain at the bottom. (We round 0.05 to a 0.)

Likewise, *always* and *never* are apparently crisp. Yet a football fan might say, "Joe Montana always comes through in the clutch," knowing he occasionally hasn't. The fan is rounding off. And if someone replies, "Wait a minute! I recall a game back in 1985 where he didn't come through," the comment is momentarily stunning. The critic is too literal, pedantic, out of touch. People understand rounding off.

As fuzzy values approach the center, rounding off becomes much less natural. For instance, each week film critics Gene Siskel and Roger Ebert describe a movie's merits, yielding a fuzzy sliding-scale evaluation, then round off to thumbs-up or thumbs-down, a crisp assessment. They can thus agree almost exactly on a film, yet round off to different poles. Since this act exaggerates their differences, they then often start squabbling.

Rounding off creates classes: *movies worth seeing, movies not worth seeing*. Likewise, *chair* includes all items we round off to it, and its definition marks the point where we stop rounding off. Such classmaking allows mental shortcuts. Except in special cases, no reasonable person tries to express the exact amount of water in a glass. The labor outweighs the prize: The imprecision of *full* just doesn't matter. The brain seeks economy, and a hazy *full* conveys all the information we need.

Moreover, as Nietzsche observed, rounding off likely reflects strong evolutionary pressures. The cat that rounds the similar mouse off to the identical one has greater odds of surviving than the cat that sees similar as only similar. Indeed, Nietzsche noted, this tendency "to treat as equal what is merely similar—an illogical tendency, for nothing is really equal—is what first created any basis for logic."[24]

Classical logic rounds off every statement into one of two classes. Hence, contradictions began arising from Cantor's sets at the turn of the century, heralded by the advent of a fictional barber. Bertrand Russell had taken up set theory and sought to prove that all mathematics was based on logic. In the course of this remarkable effort, he uncovered a conundrum:

> The barber shaves all men and only those men in town who don't shave themselves. Who shaves the barber?

If he shaves himself, he can't; if he doesn't, he must. The paradox may seem trivial, but it mirrored a grave formal problem, one involving sets that contain themselves. Consider the set of lakes. It is not a lake and does not contain itself. But a set of sets is also a set, and so can belong to itself. For instance, the set of all sets with over five members contains itself. Russell's puzzle was this:

> A set contains all sets and only those sets that don't contain themselves. Does it contain itself?

It can't contain itself, yet it must. Intriguingly, like the paradox of the heap, this puzzle too went back to the ancient Greeks, and has appeared in various other guises:

- All Cretans always lie and a Cretan says, "I am lying." True or false?
- "This sentence is false." True or false?
- "The next sentence is false. The previous sentence is true." Is either sentence true or false?

- On May 4, 1934, said Kurt Gödel, a man utters a single sentence: "Every statement that I make today is false." Is the sentence true or false?
- Consider adjectives that apply to themselves, suggested mathematicians Kurt Grelling (1886–1941) and Leonard Nelson (1882–1927) in 1908. For instance, *short* describes itself while *long* does not. *Short* is short. Likewise, *polysyllabic* describes itself, *monosyllabic* does not. *English* does, *Chinese* does not. Suppose we call the self-describing adjectives *autological* and the non–self-describing adjectives *heterological*. Does *heterological* describe itself?

This question caused much hubbub, and Russell proffered a somewhat unsatisfactory solution of it. The answer to all these questions is that they are equally true and false. They fall on the midpoint, at 0.5, equidistant from 1 and 0. We can't round off, yet we must. And that is the underlying paradox.

Crispness compels such rounding off, and it has led to many other snags in intellectual history. Two examples stand out in particular: Plato's theory of Ideals and objections to Darwin's theory of natural selection.

Fuzziness underlies perhaps the most famous philosophical scheme ever devised: Plato's theory of the Ideals.[25] Plato (427–347 B.C.) saw degrees of truth everywhere and recoiled from them. For instance, he realized, no chair is perfect. It is only a chair to a certain degree. The whole physical world comes in similar grades of imperfection—shops, bridges, clouds, smiles, paintings, clever, gentle, fascinating, big, wide, long, everything.

If an item is partly a chair, he reasoned, it is partly not a chair. It is a chair and not a chair. But that's a contradiction. Could a contradiction exist? He dismissed the notion out of hand and thus faced a dilemma. Contradiction surrounded him like the sea surrounds a fish, yet it was impossible. He resolved the problem by declaring the physical world an illusion. The floor, the lawn, the sky, the book in your hands—all are a vast mirage.

But then what was real? To answer this question, he brought forth the Ideals. Instead of this jewelry shop or that vegetable shop, he said, there was an Ideal Shop, and likewise an Ideal Bridge, an Ideal Cloud, an Ideal Chair. The Ideal Chair, in effect, was a chair to 1.0 extent. It was perfect. Moreover, all these Ideals existed in our minds from birth, and we accessed them by thought alone. Experience was delusion, but the Ideals were eternal and changeless, the only sure knowledge available.

The theory of Ideals begs so many questions that, for most modern philosophers, it is now at best a shorthand reference for other notions. We need not mesh ourselves further in it, except to note two points. First, Plato confused partial contradictions with total ones, viewing the harmony between *partly tall* and *partly short* as conflict between *tall* and *short*. This error drove him to invent the Ideals. Second, he expelled fuzziness from existence, and in so doing he vaporized the world.

Plato's Ideals were not mere intellectual zircons. Like Aristotle's axioms, they had deep, primary effects. For instance, they led to essentialism, one of the main barriers to the theory of evolution. Essentialism held that each species is an island, fully distinct from every other and based on an eternal, static essence. Variations were the imperfect realizations of the underlying model. This notion placed permanent gaps between every species and greatly hindered pre-Darwinian biologists from deducing natural selection. If a species cleaves to a timeless essence, how can it evolve?

Essentialism had the barnacle tenacity of most unseen assumptions, and according to Ernst Mayr, it "dominated the thinking of the western world to a degree that is still not yet fully appreciated by the historians of ideas."[26] In *On the Origin of Species*, Charles Darwin (1809–1882) rejected it outright. Instead, he proposed that species change over time, gradually. The Galápagos finches diversified in tiny steps over millennia, he said, rather than in abrupt jumps. There could thus be no sharp dividing line among many species. Indeed, even defining *species* was a hopeless task. "No one definition has as yet satisfied all naturalists; yet every naturalist knows vaguely what he means when he speaks of a species," Darwin wrote.[27] Taxonomists were "trying to define the undefinable."[28]

The quest to define *species* continues, and the best-known definitions, in order of general acceptance, stress reproductive isolation, common ancestry, and physical resemblance. But most biologists today recognize not only the fuzziness of species—among the 607 species of bird in North America, for instance, 46 have populations midway between subspecies and new species—but also that this fuzziness is a key to evolution.

3

THE ARAGO FACTOR

Those who have not a thorough
insight into both the signification
and purpose of words, will be
under chances, amounting almost to certainty, of reasoning or
inferring incorrectly.

—JOHN STUART MILL

Fuzzy sets . . . bring the reasoning used by computers closer to
that used by people.

—LOTFI ZADEH

■

Zadeh remained head of the UC Berkeley electrical engineering department till 1968, then decided to redirect his career. He spent half of the next year at IBM and half at MIT. IBM did no more than nod politely at fuzzy logic—"Um hmh. Very nice, very nice"—but Zadeh admits the field was still new and lacked useful engineering tools. Then, too, he did little campaigning for it, since he was busy becoming a computer person. "I was learning, learning, learning, learning," he says.

When he returned from leave, he abandoned systems and began teaching only computer courses. "I changed completely," he says. "And that had a certain impact on my work on fuzzy sets." He now saw its relevance to new problems, especially those involving language. Soon paper after paper issued from his pen, and he filled in large portions of the domain.

One of the most important innovations was hedges.

In 1971, Zadeh met George Lakoff, a colorful professor in the Berkeley linguistics department. "We started talking about this and that, and the result was two papers were written," Zadeh recalls. "One by Lakoff and one by me." Both dealt with hedges.

Hedges are terms that modify other fuzzy sets. They include *very, somewhat, quite, more or less, slightly,* and other words that alter a set's range. They lend themselves to a wide variety of propositions. For instance, they could be

- All-purpose modifiers, such as *very, quite,* or *extremely.*
- Truth-values, such as *quite true* or *mostly false.*
- Probabilities, such as *likely* or *not very likely.*
- Quantifiers, such as *most, several,* or *few.*
- Possibilities, such as *almost impossible* or *quite possible.*

These terms often qualify as operations themselves, like intersection. For instance, *very* performs concentration. It narrows a set down or, in a graph, shifts it toward the end. It creates a subset. From the set of *tall men,* it yields the subset of *very tall men. Extremely* executes the same feat to greater extent.

More or less performs dilation, the opposite of concentration. Dilation expands the set. The set of *more or less tall people* is broader than the set of *tall people*.

Some hedges lack simple English exemplars. One is *contrast intensification*, a term Zadeh borrowed from his hobby of photography,[1] where it refers to the darkroom trick of making dark hues darker and light ones lighter. In fuzzy logic, it makes true statements truer and false ones falser. Another is *contrast diffusion*, which does the reverse.

Hedges are useful as operations, but they have another virtue. They can break down continuums into fuzzy chunks. For instance, one could place the following hedges on a scale of room temperature:

Very cold
Moderately cold
Slightly cold
Neutral
Slightly hot
Moderately hot
Very hot

These fuzzy sets overlap. For instance, a temperature of 75°F does not slot fully into either *slightly hot* or *moderately hot*, but rather has memberships in both. Mapped out on graphs, each hedge is typically a triangle, bell, or trapezoid, and together they form a schematic mountain range: a series of overlapping Tetons, Half Domes, or mesas.

Hedges reflect human thought, since people do not distinguish sharply between *slightly hot* and *moderately hot*. But hedges also conduce to an artificial precision which is quite nonhuman, yet might still have uses. For instance, Zadeh said, one could define *very* as a mathematical square. Hence, if John had a 0.8 membership in the set of *old people*, he would have a 0.64 membership in the set of *very old people* and 0.4096 (0.8 to the fourth) membership in the set of *very, very old people*.

Though hedges are vague, they hardly float adrift. In 1944 Ray Simpson was worried about the elasticity of words. Why can't we be more precise? he wondered. Even scientific questionnaires abounded with such ill-defined terms as *generally* and *rather often*. To help quantify them, he asked 355 high school and college students to place 20 frequency terms like *often* on a scale between 1 and 100. How many times out of 100 was *often*?

In 1968 Milton Hakel replicated his experiment. The two results appear below, along with the median of the frequency estimates:[2]

SIMPSON (1944)		HAKEL (1968)	
WORD	**MEDIAN**	**WORD**	**MEDIAN**
Always	99	Always	100
Very often	88	Very often	87
Usually	85	Usually	79
Often	78	Often	74
Generally	78	Rather often	74
Frequently	73	Frequently	72
Rather often	65	Generally	72
About as often as not	50	About as often as not	50
Now and then	20	Now and then	34
Sometimes	20	Sometimes	29
Occasionally	20	Occasionally	28
Once in a while	15	Once in a while	22
Not often	13	Not often	16
Usually not	10	Usually not	16
Seldom	10	Seldom	9
Hardly ever	7	Hardly ever	8
Very seldom	6	Very seldom	7
Rarely	5	Rarely	5
Almost never	3	Almost never	2
Never	0	Never	0

The absolute values of these medians often varied widely but, as Hakel noted, their rank order correlated to .99.

One of the glories of the computer chip was its near-universal applicability. It could go in computers but also in blenders, automatic transmissions, light switches, lawn mowers, anywhere. Fuzzy logic has similar versatility. "You can take anything in any field and fuzzify it," Zadeh says. It is true. It is one of the remarkable features of the theory.

For instance, fuzzy arithmetic is arithmetic with fuzzy numbers. A fuzzy number is an approximate one, like *more or less* 13 and *around* 27. We encounter such figures constantly, often rounded off without comment. The population of the earth is 5 billion. The earth is 93 million miles from the sun. These are measurements, and as engineers know, all measurements err to some degree. One's height and weight, the mileage from West Palm Beach to Miami, the winning time in the Boston Marathon—all are ultimately fuzzy. When the exactness ceases to matter, we round off.

However, like gatecrashers at an inaugural ball, some fuzzy numbers present themselves in the formal evening dress of precision. For instance, one source gives the area of the Caspian Sea as 152,239 square miles. Such specificity masks an approximation, not just because of measurement error, but because marshlands are fuzzy—part land, part sea. The real size is *about* 152,239 square miles. Likewise, Simpson would have done better to quantify hedges with fuzzy numbers, for clearly whatever *seldom* means, it is not *10 times out of 100*. Fuzzy numbers make the fact explicit.

Fuzzy probabilities are a common kind of fuzzy number. When a TV weather forecaster predicts a 30 percent chance of rain, the probability is fuzzy. It means *around* 30 percent. Classic probability deals with such estimates by setting a crisp margin of error, such as plus or minus 5 percent, but fuzzy probability blurs this range. Zadeh notes that fuzzy logic can also render such likelihoods by linguistic variables such as *fairly low* instead of 30 percent.

Fuzzy geometry smudges shapes. For instance, a graph can have fuzzy points. They look like tiny clouds around a dot and represent a vague value. Likewise, an approximately straight line can wobble a bit, but not beyond a certain distance from the ideal. Fuzzy geometry permits reasoning in a near-Euclidean manner from axioms to fuzzy theorems, which are approximately true. Likewise, scholars have also probed fuzzy topologies. Topology is the mathematics of surfaces as they stretch and curve, and fuzzy topology involves approximate surfaces.

Finally, one can fuzzify fuzziness. Indeed, aspects of classic fuzzy theory were crisp to begin with. While a set may be fuzzy, a membership value like 0.7 is sharp. Yet most people do not assign a 0.7 to "Tom is tall," but a vaguer value like *fairly true*. In 1973 Zadeh suggested second-order fuzzy sets, which involve fuzziness about fuzziness. Second-order or ultrafuzzy sets have values like *about* 0.7. Like hedges, each membership is itself a fuzzy set. Theorists have gone on to develop higher orders of fuzziness.

Zadeh always stressed that he had not chiseled his classic fuzzy operations in granite. As multivalued logic could come in diverse forms, so could fuzzy operations. For instance, an intersection need not necessarily be the minimum of the memberships. It could be another amount, perhaps determined by an equation. Indeed, scholars have gone on to devise a panoply of new intersections and unions, some quite useful, like the formulas called *t-norms* and *t-conorms*. In fact, the variety of fuzzy operations is theoretically infinite. Hence, fuzzy logic is not really a single construct. It is a family.

"Sometimes there is something that sits on your nose and because it's on your nose you don't see it," Zadeh says. "That's why at first I didn't quite see this possibility theory. But *sometimes* something clicks in your mind and you begin to see that certain things fit together." He first described possibility theory in the premiere issue of *Fuzzy Sets and Systems*, the main journal in the field, in 1978. Possibility is a distinct theory from fuzziness, he notes, though possibility distribution is basically the same as membership function.[3]

What is *possibility?* The term needs explaining, as people inevitably confuse it with probability. We commonly speak of "the possibility of Uncle Jake coming to dinner" or "the possibility of Chicago winning the Super Bowl" as a likelihood. Zadeh, however, had quite a different notion in mind.

He defined possibility as the degree of ease with which an event may occur. Possibility is whether it can happen, probability is whether it will. For instance, the possibility of Hans eating three eggs for breakfast is 1.0. The probability of him doing so might be much lower, say, 0.1. Across the whole range, the two might compare as follows:

Eggs	1	2	3	4	5	6	7	8
Possibility	1	1	1	1	0.8	0.6	0.4	0.2
Probability	0.1	0.8	0.1	0	0	0	0	0

Or, in George Klir's example, probability would refer to the number of people expected to ride in a car on a certain day, possibility to the number that could ride in it at any one time.

Zadeh took pains to note that "degree of ease" was just a figurative usage. He conceived of an elastic envelope around each fuzzy set, a fuzzy restriction, which more basically indicated possibility. For instance, consider "John is young." If John is young, what is the possibility that he is 17? It might be, say, 0.6. This figure would describe how easily one could assign 17 to John, and it would depend on the elasticity of *young.*

Zadeh noted that, while reducing the possibility of an event normally reduces its probability, the reverse does not hold. Thus, if an event is impossible, it will have zero probability. However, if an event is improbable, it could be 100 percent possible, as in the case of Hans eating three eggs. He also observed that we much more often know the possibility of an event than the probability. We have a better sense of how many eggs Hans can eat than how many he will.

How do people reason? What is the secret behind the amazing human ability to reach valid conclusions? Many logicians, such as George Boole and the Port Royal Jansenists, have sought the "laws of human thought." But all reduced thinking to classical logic and truncated words into well-bounded symbols, without the everyday halo of vagueness.

In the late 19th century, the stern Gottlob Frege (1848–1925) jettisoned this goal. Critics had charged that logic was unnatural because it failed to capture words and human reasoning. Fine, he said. If the slipperiness of words mocked logic, then make symbols crisp and *meaningless*, and forget about words. The only meaning of logic, he declared, lay in the rules for manipulating symbols. The only restraint, he said, was consistency. Logic referred to nothing but itself. The symbolic logic Frege created became an imaginary palace of reason.

Zadeh took the opposite approach. Frege was chasing barren, phantom beauty, he said. We shouldn't fit words to logic, but logic to words. In 1973 he stated that fuzzy sets are the key elements in thinking. "Indeed," he wrote, "the pervasiveness of fuzziness in human thought processes suggests that much of the logic behind human reasoning is not the traditional two-valued or even multivalued logic, but a logic with fuzzy truths, fuzzy connectives, and fuzzy rules of inference."[4] It was a major step toward reconciling logic with thought.

In 1975, he spelled out fuzzy logic as logic.[5] It rests on the infinite-valued structure of Łukasiewicz, yet it is quite distinct. It is really a logic of fuzzy sets.

For instance, Zadeh said, one could have fuzzy syllogisms, such as

Most men are vain.
Socrates is a man.
Thus, it is likely that Socrates is vain.

This kind of reasoning parallels real-life thought patterns much better than crisp reasoning does. We are surrounded by fuzzy facts—"Most men are vain," not "All men are vain"—so we constantly form fuzzy judgments, such as "It is likely that Socrates is vain." Moreover, we do not sit and puzzle these conclusions out. They take shape in the mind unbidden—unlike those in classical logic, which demand special skill and training.

Zadeh suggested other forms, such as

> This stream is narrow.
> This stream and that stream are approximately equal.
> Therefore, that stream is more or less narrow.

Fuzzy logic has three unusual characteristics, Zadeh said:

- Fuzzy truth-values are words, not numbers. They include such terms as *very true, rather true*, and *not very false*. Each of these truth-values is a fuzzy set along the line from 0 to 1.
- Fuzzy truth tables are imprecise.
- The validity of fuzzy rules of inference is approximate, not exact.

Why use this approach? Zadeh asked. There were two reasons. First, by placing overlapping hedges like *rather true* and *very true* on the line from 0 to 1, an infinite set became a countable one. Fuzzy logic thus simplified the continuum. Second, it better matched human reasoning. People think in terms of "Holmes is very likely to be correct," not "Holmes is 87 percent likely to be correct," and they reach conclusions on that basis too.

Zadeh went on to devise a formal equivalent to the two-valued predicate calculus. Predicate calculus is the apex of binary logic. First described by David Hilbert and Wilhelm Ackermann in 1928, it attempts to translate natural human utterances into logical statements. It is a formal language that addresses the intent of the speaker, not the literal meaning of the words. Zadeh called his fuzzy formal language PRUF (Possibilistic Relational Universal Fuzzy). One could translate natural language expressions into it, perform inferences, then translate back into normal speech.

*   *   *

Geometry has no lock on fuzzy theorems. Such imprecisely worded proofs can expand the perimeter of science, Zadeh feels, for theorems are its girderwork. The work of a scientist, especially one who uses mathematics, "is not viewed as complete until that person can come up with a theorem," he says. "When you prove a theorem, you are establishing in a rigorous way that something is true."

Fuzzy theorems can firm up knowledge that otherwise drifts about as common sense. For instance, he says, though no one has proved a theorem in neural networks, "people can still say, 'Our experience has been, if you use this technique, you're not too far off.' It's a fuzzy theorem. It's a fuzzy statement. And my point is there are many complex problems such that the aim of coming up with a crisp theorem is unrealistic. You cannot do it. You have to settle for something less and that is a fuzzy theorem."

Zadeh notes that he has defined the concept, offered a few examples, and stopped. "But if somebody succeeded in pushing this further and showed how it can be applied in certain situations, then that would be a very significant advance, because that would open up all kinds of new directions. But that is yet to come."

TO SHIFT A PARADIGM

Zadeh might spin out the theory into an intricate tapestry, but creativity alone was not enough. As Kuhn wrote, "Novelty for its own sake is not a desideratum in the sciences as it is in so many other creative fields."[6] To gain acceptance for fuzzy logic, Zadeh had to don helmet and breastplate, confront its foes, display its worth, and win partisans. He had to unseat a paradigm.

The task can be as daunting as any in the realm of ideas. How does one paradigm overthrow another? Kuhn realized this question was a new and important one. He tentatively suggested that, though social factors such as advocates' reputation play a role, a paradigm ultimately triumphs by its elegance, consistency, comprehensiveness, and above all usefulness. However, the transition is not quite so rational, and personalities and politics play a key role in it. Like other people, scientists can follow trends, join factions with social bonds, or heed self-interest, and may try to boost or thwart a new consensus from nonscientific motives. In one fascinating study, Eugene Frankel showed how this politics of truth can work.[7]

For centuries, physicists debated whether light was a particle, as Newton thought, or a wave, as Christian Huygens (1629–1695) believed. Today, we view light as intermediate, as wavelike and particlelike. It was a lying Cretan problem, with both sides equally correct. Yet between 1815 and 1825, French physicists largely forsook particle theory and embraced waves, even though particle theory had achieved increasing successes in the period before 1815. There was no Kuhnian crisis of cognition at all. Moreover, particle advocates enjoyed total control of the research, publishing, and educational institutions in Paris, and they suppressed wave theory. What happened?

The Englishman Thomas Young (1773–1829) had advocated wave theory in the early 19th century, and by 1813 it had privately converted the powerful Dominique François Arago (1786–1853). Arago was a member of the Societe D'Arcueil, a circle of the best young physicists and chemists in Paris, which included particle advocate Jean-Baptiste Biot. As it happened, Arago was feuding with Biot over claims of priority for certain experiments.

Meanwhile, an obscure engineer in the provinces named Augustin Fresnel (1788–1827), self-taught in physics and ignorant of the recent triumphs of particle theory, undertook to prove wave theory. He studied diffraction, in which light passing through slits yields alternating bright and dark bands. Particle theory struggles to explain this pattern; wave theory requires it. Biot had downplayed its importance, and alone, Fresnel could not have provoked a crisis. But in 1815 he met in Paris with Arago. The two formed an alliance and triggered one together. By 1823, after publishing coherent explanations, clarifying theoretical mysteries, winning a major prize, and placing supporters in the Académie and at the head of key journals, they prevailed utterly. It is a tale of both persuasion and control of the channels of persuasion.

Frankel extracted several lessons from it. For instance, the scientific establishment does not usually spot a crisis first. Instead, a smaller group, often far from the major universities and centers of power, points it out. The insurgents take up a new theory and try to sway the orthodox by: (1) winning influential converts; (2) clearly communicating their views; (3) solving problems, especially ones the reigning paradigm cannot, and (4) earning credit for these solutions.

Zadeh faced these challenges. Though a well-known scientist, he could scarcely relax and wait for world acclaim, nor could he establish fuzzy theory by himself. He needed to enlist prominent supporters, convey fuzziness to other scientists and society at large, and, of course, solve serious problems and win recognition.

THE SCEPTERED CENTER

Kuhn termed the shift of allegiance to a new paradigm "a conversion experience that cannot be forced."[8] In the case of fuzzy logic, it often took the form of a shock of recognition. Reading Zadeh's work, scholar after scholar thought, "This is it! This is the way I've always thought about things."

Zadeh proved adept at cultivating such interest. His warmth and decency encouraged many, as did his special attention to promising neophytes. For instance, in 1973 Ron Yager had just written his first fuzzy paper and attended a conference where Zadeh was speaking. He asked Zadeh a question from the floor and Zadeh quickly saw Yager grasped the topic. They spoke afterwards, and Yager says, "That was the beginning of what has turned out to be a pretty good friendship." At the next conference Yager himself gave a presentation and Zadeh was the first to start clapping. "He was extremely supportive, and he's always been supportive ever since."

Such tales are legion in the fuzzy community, and perhaps no one has summed up Zadeh's personal role better than fuzzy scientist Tomohiro Takagi, who says, "I think almost all fuzzy people are supported by the spirit of Lotfi Zadeh."

Zadeh also proved tireless as a speaker and promoter. "Lotfi is the kind of guy who flies to Japan for one day," says Yager. He recalls one conference in Miami where Zadeh arrived from San Francisco half an hour before his talk, spoke for 25 minutes, ate lunch, and immediately flew back.

Yet unlike Fresnel, Zadeh had trouble winning over American scholars at the center. The center is the prize, the stronghold of a discipline. When a theory earns consensus among powerful advocates at prestigious universities, it compels attention, influences other scholars, and helps define the course of research. The theory need not be fruitful and can even be a bane. Behaviorism ravaged psychology between 1930 and 1950, yet in this period it virtually became psychology.

The eminent are thus crucial to shifting a paradigm. For instance, the Russian Nikolai Lobachevsky (1793–1856) and the Hungarian János Bolyai (1802–1860) published the first works on non-Euclidean geometry in the 1820s. But their names lacked luster and mathematicians ignored it until the death of Gauss in 1855, when his posthumous papers showed he too had worked on it. Likewise, information theory met early yawns in the United States, but caught fire in the Soviet Union. Soon after, Zadeh notes, U.S. scientists decided "that if Kolmogorov says information theory is good and Kinchin says it is good, then it must be good. So they did not have an opinion of their own. They merely followed the opinion of people they had respect for."

In the United States, aside from Zadeh, fuzzy scholars held almost no chairs at top universities. They were simply absent at Harvard, Yale, Stanford, and similar institutions. They lived in the provinces, academically speaking. This fact meant barriers, for prestigious scholars play powerful institutional roles.

For instance, they review grant applications and thereby help funnel the flow of cash. They can enrich one field of inquiry, causing it to prosper, and they can impoverish an upstart rival. They also affect the climate of opinion within funding agencies themselves. Hence, fuzzy theorists feel they have received fewer grants than they deserved.

Fuzzy scholar Maria Zemankova works for the National Science Foundation (NSF), one of the premier research funders in the United States. "The academic community has been very antagonistic to fuzzy logic, and you don't want to get sneered at," she says. "If you funded fuzzy logic research, you'd attach a stigma to yourself, because nobody from good universities was proposing this kind of work. A lot of it is that people didn't want to be associated with flops. It's not necessarily that they're antagonistic. They're just very careful, to the point of being very critical."

It is hard to prove beyond quibble that funders have discriminated against fuzzy logic. Even if figures showed disproportionately low funding for fuzzy logic, Zemankova says, a hostile observer could claim such statistics proved fuzzy proposals inferior. However, since much of

American academia dismissed fuzzy logic, it would be remarkable if funding agencies had no bias against it.

Academics also referee submissions to scholarly journals. These articles are essential to build careers and convey a new paradigm, yet a single harsh evaluation can often raise enough doubt to stop publication. As a result, exponents of new ideas can have trouble publishing. Zadeh himself was lucky to find an outlet for his 1965 article.

Even when printed, an article in a smaller and less prestigious journal can fail to reach its audience. In the classic example, Gregor Mendel (1822–1884) published his findings on inheritance in 1866, but the periodical was so obscure he did not affect the tumultuous debate over Darwinism. By 1900, when biologists discovered him, only 12 articles had cited his work.

A journal can even ban a topic altogether. Stephen Rodabaugh, a professor at Youngstown State University, says that in 1982 the editorial board of *Topology and Its Applications* refused to read further submissions on fuzzy topology. He protested strongly, noting that fuzzy topology had already yielded significant results and could help solve important real-world problems such as image resolution.[9] Managing editor Jerry Vaughan, a professor at the University of North Carolina, denies an antifuzzy policy, but admits that the journal has accepted no papers on fuzzy topology since 1982. He says all submissions have been substandard.

Established scholars also directly influence university curricula and students, the very people who later make corporate decisions about technologies like fuzzy logic. Moreover, they can boost or block the careers of young instructors. Zemankova recalls one fuzzy theorist to whom another professor said, "You'll never get tenure here, no matter what you do." The theorist then tried to switch fields and when she published fewer papers for a spell, the department held it against her. Zemankova says there are other similar stories. "People just have problems. That's why they end up in these bad schools."

Abroad, some countries accepted fuzziness much faster than others, and Zadeh feels the difference reflects the obverse of the Arago Factor: an early blessing from influential scholars. Among the academics abroad who quickly lauded fuzzy logic were Andrei Kolmogorov in the Soviet Union and Grigoire Moisil in Rumania, well-known for his work in multivalued logic. Fuzzy theorist Anca Ralescu, who grew up in Transylvania and never heard of Dracula till she reached Britain ("It's an English story"), studied under Moisil, and in 1974 her husband Daniel and C. V. Negoitia issued one of the earliest fuzzy textbooks.

In France, Arnold Kaufmann published the first fuzzy text in 1973. Kaufmann flew fighter planes in World War II and later became a professor. "He's just like Lotfi, a hard worker," says fuzzy scientist Elie Sanchez, "and when he goes to a conference, he always brings new things. He's very enthusiastic and prolific." The work commenced a four-part series. "This first book was very popular," Sanchez says. "It sold incredibly well." Sanchez himself entered the field in 1973 and would go on to become a pioneer in France.

In Germany, Hans Zimmermann, a professor at the University of Aachen, came across a paper by Zadeh and Bellman in 1972 on fuzzy decision making. "I found this article extremely interesting," he says, and on his next sabbatical he flew to California to meet the authors. He soon became a major figure in the field.

Born in Berlin in 1934, Zimmermann had built radios and similar gadgets by the time he was 10. But as World War II came to an end, Germany crumbled into chaos. "I had to flee the approaching Russians," he says, "and ended up separated from my family in Frankfurt and had to work to keep alive." He later attended universities in Germany, France, and England; taught at the University of Illinois; and joined the University of Aachen in 1967.

He sparked many early fuzzy organizations. In 1975, he started the first European working group on fuzzy sets. It included such pioneers as Didier Dubois and Henri Prade, and met occasionally for symposia where members presented and discussed papers. In 1978 he cofounded *Fuzzy Sets and Systems* and became its first editor. In 1984 he helped create the International Fuzzy Systems Association (IFSA). "In June of that year, I had Americans, Europeans, and Chinese at a conference in Hawaii," he says, "and there we launched it." He became its first president, and it remains the leading fuzzy professional group.

Wang Peizhuang ("wong pay zhwong") was among the first to introduce fuzzy logic to China. Born in Huanggang on the Yangzi River in 1936, he wrote poetry and calligraphy as a boy, and later became interested in math. In 1953 he entered Beijing University, China's best, and in 1957 he graduated and became a teacher, specializing in Markov processes and probability in general. Meanwhile, the Chinese government had thrown his father, a professor and member of Chiang Kai-shek's Guomindang, in jail, where he died in 1962.

In 1966 the Cultural Revolution began, the terrible, bizarre upheaval in which thuggish Red Guards roamed the streets and Mao Zedong sent millions of intellectuals to hoe the earth. From 1967 to 1971, Wang labored first on a farm, where this natural optimist sang as he tended the sheep, and then in a motor factory, forging molten steel eight hours a day. Later, back at Beijing University, he spent six months shoveling coal in the furnace. "It helped me to see the world more thoroughly, to see not only positive but the negative," he says. "Both sides have to be considered. This was an early fuzzy experience."

The Cultural Revolution ended in 1976, the year of Mao's death. Throughout it all Wang had never entered a library, but the following year he ventured back into Beijing Library. "I was looking back at what had happened in the past ten years in mathematics," he says. There he discovered Ralescu and Negoitia's text on fuzzy sets. It impressed him at once, but since it was new, he could not check it out. So he returned to the library every day for a month, translating it into Chinese. The book mentioned Zadeh's 1965 paper, which he soon tracked down and read with appreciation. "I respected Zadeh from the first," he says.

Within two or three years, Wang and his colleagues had written many papers and teaching materials, and given seminars and courses. In 1981 he helped found the Chinese Mathematics and Fuzzy Systems Association, and today China has more fuzzy specialists than any other nation on earth.

Fuzzy logic took root early in Japan. Kokichi Tanaka of Osaka University introduced it to the island nation in the late 1960s, along with Kiyoji Asai of the Osaka Institute of Technology. Tanaka also taught Masaharu Mizumoto, who published the first fuzzy paper. Soon there were three early working groups in fuzzy logic: Tanaka's, Asai's, and Toshiro Terano's.

Noboru Wakami took Tanaka's course on information theory at Osaka University in the 1960s. Student riots often shut down classes, so Tanaka delivered only five or six lectures. It was just as well, Wakami thought. "He was very severe to the students," he says. "Most professors lectured whether students were listening or not. But Professor Tanaka became very angry if we were not listening. He was very earnest."

Toshiro Terano is a very different person, merry and affable. In the early 1970s he visited the National Cybernetics Research Institute in Naples. "This institute was very interesting," Terano recalls. "There were many, many people there—engineers, information scientists, artificial intelligence experts, bionics experts, medical doctors. They studied everything fuzzy, in the wider meaning." There he came across Zadeh's 1965 paper. He already knew Zadeh's name from system theory, but this paper surprised and interested him. He returned to Japan and started his own working group in 1971.

In 1974, Japanese and American fuzzy researchers met at Berkeley in the first U.S.–Japan Fuzzy Seminar. At the time, Terano estimates, only 10 people were studying fuzzy logic in Japan. Nonetheless, it was a key opportunity for Japanese to meet fuzzy theorists in the United States, and it yielded a collection of papers, one of the earliest on fuzzy logic in Japan. In 1980, the two working groups, Terano's and Asai's, held a joint conference in Osaka, and in 1985, they merged.

CRACKLE ON THE LINE

As fuzzy logic spread, animosity toward it grew. Remarkably, much of it stemmed from ignorance. "It's very difficult to find a person who is hostile and well-informed," says Zadeh. "So it's simply a matter of making people aware of what the issues are. Once they become aware, they change their position."

Zadeh himself was indefatigable. He wrote a wealth of lucid papers, commencing almost every one with a basic explication of the theory. Even so, the task of conveying fuzzy logic proved remarkably difficult. Fuzzy advocates had to present it clearly and vividly, and they did not always succeed.

The name *fuzzy logic* created instant communication problems. Indeed, the term has shown an uncanny tendency to shunt people into a hall of mirrors. But there were other difficulties too. Most fuzzy papers were and are highly technical, clogged with formulas. Though they were solid, only specialists could immediately grasp them. They thus forfeited the attention of the wider scientific community—especially social scientists, who often lack a mathematical bent. Even pro-fuzzy scholars of uncertainty, such as Michael Smithson, confess they have trouble following the "needlessly forbidding complexity and formality" of fuzzy articles.[10] Fuzzy logic would ultimately rise through its utility, but those best placed to use it often failed to grasp its merits.

Moreover, says Zemankova, there was a lack of good textbooks at the undergraduate level. Fine texts did exist, such as those by George Klir, Zimmermann, and Dubois and Prade, and "they could be used by a graduate student with some background, but not someone jumping into it from an entirely different field," she says. Social scientists and many engineers had to pick their way through a bramble-field of specialization.

Mathematicians had no trouble grasping the formulas, but here a different problem arose. According to Bart Kosko, fuzziness lacked a firm foundation, a structure dovetailing with other mathematics, especially probability. Indeed, to probabilists it was an isle swathed in mist, and they ignored or scorned it. In a sense, it spoke the language of neither social scientists, engineers, nor mathematicians.

The clearest argument, of course, is the solution no other paradigm can provide. Most such breakthroughs would appear in technology, but despite the obstacles, some arose early in the social sciences, where they confirmed both the usefulness and the central insight of fuzzy logic.

4

THE SHARP MIRAGE

■

Chaucer, Henry James, and, very humbly, myself, are practicing the same art. Miss Stein is not. She is outside the world-order in which words have a precise and ascertainable meaning and sentences a logical structure.

—EVELYN WAUGH

Words [resemble] those nebulous masses familiar to the astronomer, in which a clear and unmistakable nucleus shades off on all sides, through zones of decreasing brightness, to a dim marginal film that seems to end nowhere, but to lose itself imperceptibly in the surrounding darkness.

—JAMES A. H. MURRAY

Think of arm chairs and reading chairs and dining-room chairs, and kitchen chairs, chairs that pass into benches, chairs that cross the boundary and become settees, dentist's chairs, thrones, opera stalls, seats of all sorts, those miraculous fungoid growths that cumber the floor of arts and crafts exhibitions, and you will perceive what a lax bundle in fact is this simple straightforward term. In cooperation with an intelligent joiner I would undertake to defeat any definition of chair or chairishness that you gave me.

—H. G. WELLS

In the *Euthyphro*, Socrates is standing outside court when a self-proclaimed religious expert named Euthyphro happens by. They start chatting, and Socrates begs him to define the *pious*. You know so much about piety, Socrates says, that surely you can tell me what the word means. Flattered, Euthyphro offers several definitions. Socrates analyzes them into idiocy, and after much punishment Euthyphro seizes a chance to steal away.

Was Socrates being fair? Do words like *pious* have clean edges, or do they blur, as Zadeh contends?

Western thought has assumed that they are crisp. Thomas Hobbes, for instance, urged the necessity of exact definitions, and John Locke held that everyone could "strip all his terms of ambiguity and obscurity."[1] Indeed, Locke even wondered whether, given our free and easy treatment of words, language had "contributed more to the improvement or hindrance of knowledge amongst mankind."[2]

Moreover, like Descartes, Locke believed words were composed of smaller parts, like building blocks. To clarify a word, he advised its dismantling and scrutiny. Once we break *justice* into its parts and gain a "distinct comprehension" of them, he said, we will have certain grasp of its meaning. The young Leibniz attempted to discover this little "alphabet of human thought," but eventually gave up and suggested it was infinite.

A few Western philosophers balked at this approach, at least implicitly. Bishop George Berkeley (1685–1753) and David Hume (1711–1776) held that concrete ideas form the core of concepts and attract others by resemblance. We apply a word to a host of similar objects, Hume wrote, "whatever differences we may observe in the degrees of their quantity and quality."[3] Neither Berkeley nor Hume addressed fuzziness outright, though their writings plainly suggest it. Immanuel Kant (1724–1804), who read Hume carefully, declared that only mathematical quantities had clean definitions. And working lexicographers like James Murray (1837–1915), who led the group that developed the *Oxford English Dictionary*, had little patience with crispness.

But these were murmurings, and the ancient substrate, the Laws of Contradiction and the Excluded Middle, persisted almost unexamined. Words, classes, and concepts had sharp boundaries.

In this view, categories—that is, most words—have defining attributes. A word is like a Cantor set, and definitions lay out the circular bounds. Moreover, all members of the class belong to it equally. Once a word passes the checkpoint of definition, it enters a utopia of peers. A deck chair, a Chippendale, a beanbag chair—all are equally chairs.

Since the early 1970s, scientists have pummeled this view, and today, in pure form, it has about as many advocates as the Bogomile heresy.

HOW GOOD A BIRD?

The torchbearer of this revolution was Eleanor Rosch, a psychologist at UC Berkeley. She would not just argue the fuzziness of words, but prove it. Her simple 1973 experiment took only 10 minutes, yet had an impact which society has still to absorb.

She gave 113 college students lists of six words in eight categories: fruit, science, sport, bird, vehicle, crime, disease, and vegetable. On a scale of 1 to 7, she asked them to rate how well each word typified its category, how good an example of *fruit* or *bird* it was. For instance, if an ostrich were an excellent instance of *bird*, they should give it a 1, but if just a fair one, a 4.

She obtained her first finding before the students' pens touched paper: They understood these instructions at once. They neither protested nor questioned the task and, when she told them to begin, raced through it, averaging three seconds per choice. She concluded that the experiment assessed natural, everyday mental processes.

She obtained the following ratings:[4]

FRUIT		SCIENCE		SPORT	
Apple	1.3	Chemistry	1.0	Football	1.2
Plum	2.3	Botany	1.7	Hockey	1.8
Pineapple	2.3	Anatomy	1.7	Gymnastics	2.6
Strawberry	2.3	Geology	2.6	Wrestling	3.0
Fig	4.7	Sociology	4.6	Archery	3.9
Olive	6.2	History	5.9	Weightlifting	4.7

BIRD		VEHICLE		CRIME	
Robin	1.1	Car	1.0	Murder	1.0
Eagle	1.2	Scooter	2.5	Stealing	1.3
Wren	1.4	Boat	2.7	Assault	1.4
Ostrich	3.3	Tricycle	3.5	Blackmail	1.7
Chicken	3.8	Skis	5.7	Embezzling	1.8
Bat	5.8	Horse	5.9	Vagrancy	5.3

DISEASE		VEGETABLE	
Cancer	1.2	Carrot	1.1
Malaria	1.4	Asparagus	1.3
Muscular dystrophy	1.9	Celery	1.7
Measles	2.8	Onion	2.7
Rheumatism	3.5	Parsley	3.8
Cold	4.7	Pickle	4.4

Not only did these results match basic intuition, but the subjects agreed strongly among each other, especially as to the best example in each category. For instance, all 113 ranked *chemistry, murder,* and *car* as best examples.

This experiment, whose results scientists have replicated many times and which are now widely accepted, overturned the ancient tradition of thinking about human categories. It proved Zadeh's basic assertion: Classes are fuzzy.

Moreover, Rosch found borderline members seemed to cause more uncertainty. In a follow-up, she asked subjects to respond true or false to assertions such as "A carrot is a vegetable" and "A pickle is a vegetable." She found they answered true significantly faster with high-ranking items like *carrot* than low-ranking ones like *pickle.* The marginal examples demanded more thought. Like Siskel and Ebert's so-so movies, they seemed harder to round off.

In the following years, Rosch made further discoveries. She suggested that fuzziness extends from the edge of categories to the center. A robin is a better *bird* than a seagull, and a seagull is a better *bird* than a penguin. A sedan is more *car* than a station wagon, which is more so than a dune buggy or an ambulance. She called best examples like *robin* and *sedan* prototypes.

Rosch took the term *prototype* from earlier literature. It has bred some confusion, since it suggests that a prototype, like a Platonic Ideal, is somehow a model for the class. She later clarified it: "To speak of a 'prototype' at all is simply a convenient grammatical fiction; what is really referred to are judgments of degree of prototypicality. Only in some artificial categories is there by definition a literal single prototype."[5] The heart of her theory is degrees, and a prototype is merely any excellent example, whatever it may be.

In Zadeh's sense, *prototypical* performs concentration on classes, and moving from *birds* to *good birds* resembles going from *tall* to *very tall*. However, prototypes turned out to have important features.

For instance, Rosch discovered that children learned prototypes earlier than peripheral examples and she followed Berkeley and Hume in speculating that children initially think of a class in terms of concrete cases rather than defining features—sparrows and eagles, not beaks and feathers. It was further evidence that classes attract members by resemblance, by degree.

In fact, Rosch and her co-workers found that the definition of a class yields few clues about its best members. No definition of *bird* mentions wildness or tameness, yet people judge wild birds as better birds. Entering the class demands one set of features, while going to its head may require another.

So, she asked herself, what determines the best examples?

She turned to Ludwig Wittgenstein (1889–1951). In his later years, the brooding Viennese repudiated his stark *Tractatus Logico-Philosophicus* and came to believe that words referred to items without common characteristics.[6] For instance, he asked, what is a *game*? Is it a competition? But noncompetitive children's pastimes like ring-around-the-rosy are *games*. Is it an amusement? But we refer to *war games* and *deadly games*. Does it involve two or more people? But solitaire and pinball are *games*. Words like *game*, he suggested, are catchalls for items with a family resemblance. Each game may have one or more traits in common with various others, but there may be no traits common to all.

In 1975, with Carolyn Mervis, Rosch tested this idea.[7] She began with high-level categories such as *furniture* and *clothing*, comprised of concrete classes such as *sofa* and *jacket*. She presented subjects with twenty items in six categories, four of which appear below in order of prototypicality:

	FURNITURE	VEHICLE	WEAPON	CLOTHING
1	Chair	Car	Gun	Pants
2	Sofa	Truck	Knife	Shirt
3	Table	Bus	Sword	Dress
4	Dresser	Motorcycle	Bomb	Skirt

5	Desk	Train	Hand grenade	Jacket
6	Bed	Trolley car	Spear	Coat
7	Bookcase	Bicycle	Cannon	Sweater
8	Footstool	Airplane	Bow and arrow	Underpants
9	Lamp	Boat	Club	Socks
10	Piano	Tractor	Tank	Pajamas
11	Cushion	Cart	Teargas	Swimsuit
12	Mirror	Wheelchair	Whip	Shoes
13	Rug	Tank	Icepick	Vest
14	Radio	Raft	Fists	Tie
15	Stove	Sled	Rocket	Mittens
16	Clock	Horse	Poison	Hat
17	Picture	Blimp	Scissors	Apron
18	Closet	Skates	Words	Purse
19	Vase	Wheelbarrow	Foot	Wristwatch
20	Telephone	Elevator	Screwdriver	Necklace

She asked subjects to describe the features of *sword, piano,* and other words on the list. She found that the better the prototype, the more qualities it shared with other words on the scale. Sharing was central: Overlap of features made words prototypical.

Yet members of a class rarely shared one trait overall. In two categories subjects mentioned no common feature, and in the other four, only one. Even this lone attribute was doubtful in three cases, for it was an overbroad quality like "you eat it," which describes *fruit,* but also *meat, vegetables, desserts,* and all other food.

At the same time, the better the prototype, the less it had in common with contrasting classes. A sword and scissors may be equally lethal, but *sword* does not immediately suggest categories beyond *weapon*. *Scissors* does, and hence is not a prototype. Prototypes act like creosote bush in the Mojave Desert—they space themselves away from each other. They not only exemplify the class, but distinguish it.

She went on to probe the long-standing rationalist idea that words are made up of building blocks. If words are crisp, then it makes sense that they might begin at some atomic level and build up to make all the concepts and words we know. They can fit together like bricks into larger and larger units. But if they are fuzzy, this scheme collapses.

She made a remarkable finding. Categories form a hierarchy, with three vaguely bounded tiers: superordinate, basic, and subordinate.

At the top stand the superordinate categories, like *furniture*. They are abstract and no one object clearly represents them. Instead, they are collections of basic categories, such as *chair, sofa, lamp*.

The basic categories fit in between. They are fundamental. A basic category is the largest class of which we can form a fairly concrete image, like *chair*. People tend to recognize their shape and use the same motor movements with them. They are the first classifications children make.

At bottom lay subordinate categories, the divisions of basic classes, such as *kitchen chair* and *deck chair*. These tend to share most of their attributes with others in the basic category. For instance, *kitchen chair* and *deck chair* overlap much more than *chair* and *table*.

Hence, people begin forming concepts in the middle. We grasp immediate classes like *chair*, *car*, and *apple*, and both add them together synthetically to form larger categories and carve them up analytically to make smaller ones.

From an evolutionary perspective, this approach makes far more sense than the atomism of Locke. Animals need to identify concrete objects above all else. The cat must recognize the mouse at once, as a whole, because if it mentally assembles parts, the rodent escapes. A focus on Rosch's basic categories has survival value.

Family resemblances do not completely define a class. Some categories, such as certain shapes (*line* and *corner*) and facial expressions (*smile*), may involve inborn neural responses. Nor does commonality of features totally determine a prototype. Well-known words are more likely to become prototypes, as are those with certain traits, like intensity.

"I spent years trying to get my early things published," Rosch recalls. "Journals would send them back, finding fussy little things wrong with them, and saying, 'Everyone knows this isn't true.'" Why did they balk? "If something is going to be new, it has to be exactly in the right degree of difference from what's going on for people to say, 'That's interesting,'" she says. "If it's too new, it isn't understood."

She encountered similar problems with audiences. She often explained family resemblances in speeches and found that some listeners inevitably objected, insisting that classes had defining features. She would solicit such definitions and, like Socrates, vaporize each one with counterexamples. Even when unable to point to a single valid definition, people clung to this notion. Why?

When thinking of a category, we normally visualize the clear cases, not the vague ones. *Chair* evokes a dining room chair, not a beanbag chair. Perhaps, she thought, the best examples actually do share many features. She examined the top five members in each category and found that, although category members overall had few common features, the top five had many. Since people see class prototypes as the class itself, they fall prey to the illusion of common elements.

Why do we focus on prototypes? In 1978, looking back on her work, Rosch suggested that the purpose of classes was "to provide maximum information with the least cognitive effort."[8] There is a trade-off: information versus labor. And prototypes ease the burden. For instance, the eye can discriminate 7.5 million colors. If definitions were sharp, if *blue* ended at a single hue, people would need a pointless acuity of perception. One might as well demand a watch accurate to a second per century, when a much less costly one would do just as well.

Cognitive economy also likely increases the speed of the brain, enabling it to work with essences rather than time-consuming fringe items. Zadeh had observed that words summarize, and prototypes summarize further. They catch the gist of the fuzzy set.

If Rosch is correct, the mirage of words is one of the deepest and most pervasive in Western history. It helps explain why Plato thought his Ideals were possible, why Aristotle devised the Law of the Excluded Middle, why biologists fell victim to essentialism, why so many philosophers and scientists over the centuries have assumed that categories are crisp, even while acknowledging the pesky problem of vagueness. It also partly explains the hostility to fuzzy logic, since if people naturally round classes off to their core, they will tend to ignore the ever-present nimbus of gray.

Moreover, Rosch's work helps answer a further question about fuzziness: If words have such hazy borders, why don't they confuse us more often? We think of them in terms of prototypes, which are fairly clear. Scholars may expend much ink on whether a fringe item like a giant panda is a bear or raccoon, but they know what bears and raccoons are. And Euthyphro understood the *pious*, even if he couldn't define it.

THE FORMALISM AND THE MIND

"Prototypicality and fuzziness come from very different fields," Rosch says. "Zadeh has a formal definition and I have an empirical definition. I suspect they'd map onto an actual category in much the same way."

However, she never directly used fuzzy theory and in fact eschewed all formalisms. The first to apply fuzzy logic to words were Harry Hersh and Alfonso Caramazza, of Johns Hopkins University. They examined hedges.

The two showed undergraduates slides with black squares of different sizes and asked which of 13 hedged adjectives—such as *not very large*—described them. The student responses closely paralleled Zadeh's fuzzy sets. For instance, when plotted on a graph, opposites like *not small/not large* and *very very large/not very very large* formed near-mirror images.

Likewise, the union phrase *either large or small* followed the maximum of *large* or *small* and formed a gully at the midpoint of the graph. However, it was somewhat lower than these terms, and to fully represent fuzzy union, it should have equaled them. Hersh and Caramazza suggest that some subjects may have had trouble grasping the notion.

The one anomaly was *not very large*. Zadeh had defined it as everything smaller than *very large*, and the subjects so interpreted it. However, in everyday English, it has the ironic meaning *sort of small*. Hersh and Caramazza believed that the subjects may have taken cues from the matrix of logical terms. Hence, they performed another experiment, assessing responses to *large, very large, not very large,* and *sort of large*, and found people here treated *not very large* as *sort of small*. This fact, they concluded, had no impact on fuzzy theory, but added a tiny gloss on the term itself. Word usage does not always follow logic books.

Gregg Oden was a graduate student at UC San Diego when he heard linguist George Lakoff give a speech on fuzzy theory. He talked about hedges, using classic Lakoffian examples: Is Esther Williams a fish? Is she a virtual fish? Is she sort of a fish? "It just clicked," Oden says.

He began to pursue it. "Back then, in 1971–72, the literature was still manageable. The aspects of it I can apply to psychology are the most general, least technical parts." So he studied the papers in the literature, then sat down to determine how fuzzy logic could illuminate the psychology problems that interested him.

Words reflect concepts. If words are fuzzy, it suggests concepts are too. In 1977 Oden, by then a professor of psychology at the University of Wisconsin, published a study which applied fuzzy logic to semantic memory. We can recall information in many ways, such as words by their first letter or events by chronology. Semantic memory is memory by meaning. Most models of semantic memory relied on classes and supersets, and Oden realized that if these categories were fuzzy, the theories would require major alteration.

Zadeh claimed people reason in fuzzy terms. Oden sought to determine if they can actually do so, at a simple level. He presented 30 university students with pairs of statements, such as "A chair is furniture" and "An ostrich is a bird," and asked them to determine which of the two was more true and how much truer it was. The subjects responded by placing a pin in a corkboard, with its position showing degree. Because the subjects were comparing memberships in fuzzy sets, the experiment involved low-level reasoning with fuzzy information.

He found that the fuzzy model predicted the responses very closely, at both a group and individual level, and noted that this kind of information was hard to obtain in any other way. Moreover, he said, our ability to process fuzzy information should "permeate virtually every cognitive process that uses semantic information." Fuzziness suffuses meaning.

In 1978 anthropologist Willett Kempton of UC Berkeley performed an experiment that yielded a surprising result.[9] He showed subjects 51 table utensils and asked whether each was a pitcher, cup, coffee cup, mug, jar, or drinking vessel. He used both a crisp approach—is it a jar or not?—and a fuzzy approach—how much is it a jar? Subjects could choose from six grades of membership, such as *absolutely not a jar*, *not really a jar*, and *sort of a jar*.

Most expected predictions came true. He found, as others had, that the hedges made sense to the subjects and had fuzzy boundaries. *Coffee cup* was a subset of *drinking vessel* and of *cup*. *Drinking vessel* and *jar* were separate sets.

But the relation of *cup* to *mug* proved odder. The crisp questions showed that *mug* was a simple subset of *cup*. However, the fuzzy questions revealed that some objects were more *cup* than *mug*, as one would expect, but also that others were more *mug* than *cup*. Part of *mug* was a subset of *cup*, yet at the same time, it appeared, part of *cup* was a subset of *mug*.

Fuzzy theory could not describe this phenomenon, Kempton noted, since Zadeh had defined fuzzy subsets absolutely. A fuzzy set is either a subset or it isn't. "However," Kempton asked, "if elements can be members of sets to a degree, we might ask if it could also be true that sets are subsets of other sets to a degree."[10] Suppose *mug* were a 0.75 subset of *cup*. Subjects would then round off and crisply describe a mug as a kind of cup, without thinking every mug was a kind of cup.

The difference, Kempton noted, mirrored a simple shift of the indefinite article in English. "A mug is *a kind of* cup" or "*a sort of* cup" means that *mug* falls wholly within *cup*. But "A mug is *kind of a* cup" or "*sort of a* cup" means that *mug* and *cup* are partial subsets of each other.

Crisp sets also overlap in degrees. However, he noted, crisp taxonomic theory offered no way to measure such differences in relationship, much less explain them. But an extended fuzzy theory, he concluded, offered a powerful formalism for modeling cognition.

THE MYSTERIES OF COLOR

Biological systems were very much on Zadeh's mind when he invented fuzzy logic, and in fact it may explain how we see color—not just how we name it, but how the mind identifies colors themselves.

Color was an early area for pioneers in categorization, since they believed color names were arbitrary. They saw the spectrum as a smooth, evenly changing band that cultures partitioned at whim. Indeed, anthropologists had found that languages vary remarkably in the number of terms they have for colors. Hence, they felt it was a good arena to test the famous Sapir–Whorf hypothesis, that words shape perceptions.

But are color names arbitrary? In 1969, Brent Berlin and Paul Kay of UC Berkeley discovered that people from diverse cultures readily agree on what makes a "good green" or a "good blue," even if they have no name for it. They have a harder time with intermediate colors like chartreuse. Moreover, cultures showed a fixed sequence in naming colors. If, like the stone-age Dani of New Guinea, they possess only two words for color, these refer to black/cool and white/warm. If a third appears, it denotes red. The fourth and fifth are yellow and green (in either order). The fifth and sixth are blue and brown (again in either order). The last four are purple, pink, orange, and gray. Not only are color names not arbitrary, but they seem predetermined.

Paul Kay had worked with Zadeh in a Berkeley study group, and in 1978 he and Chad McDaniel proposed that the neurophysiological mechanism beneath colors is continuous, and that fuzzy logic best models it.[11]

How does the eye see color? Myths about this issue abound. For instance, most people think red indicates the longer wavelengths of light and hence appears on just one side of the spectrum. However, it also surfaces slightly on the violet side, where it helps form purple. But if red is the longer wavelengths of light, how can it?

In the early 19th century, Thomas Young suggested the eye forms all colors from three primaries (red, yellow, and blue). Hermann von Helmholtz (1821–1894) later elaborated the notion, holding that three different kinds of nerve (now red, green, and blue) conduct color sensations to the brain, which derives the remaining colors. This simple scheme has appeared in most discussions of the topic and dominated popular thinking on color.

However, a second, less publicized approach has paralleled it all along. Johann Wolfgang von Goethe (1749–1832) argued rather hotly that there were six basic colors: red, green, blue, yellow, black, and white. The German physiologist Ewald Hering (1834–1918) investigated more deeply and scientifically and found surprising phenomena. For instance, white is not necessarily the presence of all colors at once. It can arise from mixing blue and yellow alone. Likewise, if you stare at a red light and suddenly shut it off, a green afterimage appears, and vice versa. Place red against a gray background, and the eye sees green at the boundary between red and gray. The color blind, he found, can usually see neither red nor green. Moreover, while many kinds of color can fuse to form new colors—red and blue, for instance, make purple—red cannot mix with green, nor blue with yellow. He thus theorized that color derives from basic oppositions: red and green, and blue and yellow, as well as black and white.

Thirty years ago Hering's theory was almost forgotten. Today scientists know he was right. The retina senses color through cells called *cones*. There are indeed three kinds of cone, and they respond differently to wavelengths of light, peaking at violet (419 nm), green (531 nm), and yellowish-green (558 nm). But cones are just the atrium of visual processing. They don't send signals straight to the brain, but rather along neural pathways to another kind of cell, a middleman called an *opponent response cell*. These cells remake the message into terms the brain will read.

Opponent response cells come in four types. Each fires spontaneously at a certain rate in the absence of any stimulus. It is like a normal resting heartbeat. Like a heartbeat too, it can speed up or slow down. A change in either direction conveys color information to the brain.

The four cells come in two pairs, and each cell mirrors its partner. In the red–green pair, one cell fires more rapidly with red and more slowly with green, while the other does the opposite. In the yellow–blue pair, one cell fires more rapidly with yellow, more slowly with blue; its twin reverses the pattern. The overall firing rates of these cells determine the colors we see. Hence, red can appear at the violet end of the spectrum because it is not the longer wavelengths of light. It is the ticking of a cell.

These hues are not all, however. The eye also registers black and white. These are not opponents like blue and yellow, so we can perceive them simultaneously in the visual field, where they mix to form gray. Ironically, black and white, the epitome of duality, easily and gently flow into each other. The real duality lies in blue and yellow, red and green. Like day and night, red and green cannot tolerate each other's presence, yet they are inseparable.

Kay and McDaniel stated that each of the four colors forms a fuzzy set, with membership peaking at a focal color and declining in either direction to zero. Fuzzy operations like union and intersection could then explain every color term in every language of the world.

Languages in early stages of evolution, lacking enough words to describe each basic color category, use composite terms like, say, *grue* for "green and blue." *Grue*, they said, is a fuzzy union. It included any color with membership in green OR blue. It thus forms the classic camelback profile, and in anthropological studies, subjects always give bluish-green less membership in *grue* than either blue or green.

A total of 57 fuzzy unions are possible with the six primaries, but anthropologists have seen only three others: "warm" (red OR yellow), "light-warm" (white OR red OR yellow), and "dark-cool" (black OR green OR blue).

Once cultures have named all the basic primaries, they begin naming colors in the gaps—brown, orange, pink, purple, and gray. These new labels are fuzzy intersections. For instance, green and yellow intersect to yield greenish-yellow, or chartreuse. It includes, with varying degrees of membership, colors from pure green to pure yellow. Likewise, brown is the overlap of yellow and black, pink of red and white, purple of red and blue, orange of red and yellow, and gray of black and white.

However, here a problem arose. The absolute values of an intersection like *orange* did not reflect experience. It peaked around 0.5 and thus suggested no good examples of orange existed. In fact, subjects were just as confident in assigning a good orange to *orange* as a good red to *red*, and many said that, for them, orange was just as distinctive and fundamental. Yet fuzzy intersection implied that no color could be more orange than it was yellow or red.

So Kay and McDaniel modified the model, deriving a workable *orange* by simply doubling the value of the intersection. They "normalized" it, that is, stretched it till it covered a scale from 0 to 1.0. As a result, focal orange rose to as high a peak as focal red or yellow.

On the other hand, chartreuse differs from orange and the other five intermediate colors. It does not seem to be a basic category, and Kay and McDaniel speculated that it might behave as a simple intersection.

Does the brain use fuzzy logic with color? Kay and McDaniel reserved judgment, but suggested that it likely takes some similar approach, since subjects need slightly longer to identify intermediate colors than primary ones. "The formalism of fuzzy set theory," they concluded, "provides a natural framework for the description of these universal and non-discrete categories."[12]

In 1983, Don Burgess and co-workers investigated color terms among the Tarahumara Indians of Chihuahua in northern Mexico. The Tarahumara tongue is unusual in that it has adverbs indicating how much a color term applies, and speakers must add these adverbs every time they mention a color. They can't say "blue" but must qualify it as "medium blue" or "light blue." The investigators concluded that these modifiers followed the rules of fuzzy sets, and their data strongly supported Kay and McDaniel's model overall.

BRIDGING WORDS AND NUMBERS

"The social sciences are legendary for the arguments between proponents of qualitative and quantitative research styles," says Michael

Smithson, a professor at James Cook University in Australia.[13] Because fuzzy sets model words mathematically, they map the numerical onto the verbal and can bring these two worlds into sync. Thus, Zadeh predicted, fuzzy logic can play a unifying role in the social sciences.

For instance, in economics the mathematical and linguistic realms stand quite apart. Since economics is money and money is numerical, math brings powerful tools to the field. Yet the precision of math leads to overly crisp estimates and idealized models that seem to describe a society of robots. Hence, economics also employs verbal concepts like recession. Language handles real-life questions better and treats details more subtly, but it also narrows the scope of models and shortens chains of reasoning.

In 1981 Arne Pfeilsticker of the University of Heidelberg suggested that fuzzy sets could end the no-man's-land between these two and bring them into alignment. Economics, he said, abounds with fuzzy notions. Is a person who works four hours a day but wants to work eight *unemployed*? Partly. The notion is fuzzy. Likewise, gross national product (GNP), inflation, investment, households, market-dominating companies, and many other concepts were also fuzzy.

Thus if an economist defined *recession* by the number of months of falling GNP, she could create a classic fuzzy set. The more months, the more recession. By merging curves for other factors, such as months of declining income and increasing unemployment, she could develop a more sensitive measure. In any case, the mathematical result would be more accurate and it would spare everyone the comedy of respected authorities arguing about whether we are "in" a recession or not.

By now a variety of fuzzy economic models have appeared, such as those of the theory of value, and in 1988 Claude Ponsard of the Institut de Mathématiques Economiques in Dijon, France, declared that fuzzy theory had led to economic insights unobtainable by classical methods.

Fuzzy logic lends itself to other social sciences in much the same way. "Translation of even simple hypotheses into mathematical form has been notoriously slipshod in the social sciences," said Smithson, and is "one of the most frequently voiced criticisms of quantitative methods."[14] Fuzzy sets could minimize this garbling and lead to more acceptable and illuminating use of math.

Moreover, he said, fuzzy sets clarified thought and aided the building of theories. Fuzzy tools of analysis could spur more research into reasonable but fuzzy theses such as "a high level of education comes close to being a necessary condition for democracy." In particular, he felt possibility theory could keep social scientists from confusing notions like choice and preference with odds and probability.

Despite these prospective boons, fuzzy logic has not yet made deep inroads in the social sciences. Why not? Smithson suggested several reasons. In fields such as psychology and sociology, he said, the methodologically innovative tend to know little math and the mathematically astute tend to shun novel methods. Moreover, few scholars had attempted to show the relevance of fuzzy logic to the social sciences. "Crudely put," he said, "fuzzy set theory has not been presented in a user-friendly manner."[15]

And then there were Osherson and Smith.

PET FISH AND LINDA

In 1981 psychologists Daniel Osherson of MIT and Edward Smith of Stanford published a jaunty critique of prototype theory, and fuzzy theory as well, concluding that the entire prototype–fuzzy effort was a carnival of contradiction.

They first assailed fuzzy logic for breaking the Laws of Contradiction and the Excluded Middle, which they felt were intuitively true. There could be no "apple that is not an apple," they said, so all membership values in this illusory set must be zero. Yet in fuzzy theory a misshapen real apple could have a 0.9 membership in *apples* and a 0.1 membership in *not-apples*. Its membership in *apple and not-apple* would thus be 0.1, and we would be gazing at an apple that is not an apple. In a similar way, they showed, fuzzy sets violate the Law of the Excluded Middle.

George Lakoff disposed of this ancient argument quickly, noting that a carved wooden apple might be an apple that is not an apple, as might a cross between an apple and a pear. "Osherson and Smith do not consider such possibilities," he said.[16] Once again, intelligent analysts had construed partial contradiction as total contradiction.

But the two critics found graver problems with fuzzy intersection, which recall the difficulty with *orange*. Consider *pet fish*. It is an intersection, the common ground of *pets* and *fish*. So how much is a guppy a *pet fish?* It must be the lesser of its memberships in *pets* and *fish*. But this makes no sense, they argued. A guppy is plainly not prototypical of *pets*. It is a fairly uncommon pet. Nor is it prototypical of *fish*. Both of these memberships would be low. Yet the guppy's membership in *pet fish* would be high. Intuitively, it is more a *pet fish* than either a *pet* or a *fish*. Likewise, they noted, a striped apple is more prototypical of *striped apple* than either *striped things* or *apples*. Yet by Zadeh's rule of intersection, it had to be less. The fuzzy claim to model human thought explodes.

Or does it? As it happened, Osherson and Smith had touched on the intriguing cognitive puzzle called the conjunction fallacy. In a famous experiment, psychologists Amos Tversky and Daniel Kahneman presented subjects with a description of "Linda," mentioning that she was 31, outspoken, and concerned about discrimination and social justice. They then asked whether it was more likely that Linda was (1) a bank teller, or (2) a bank teller and a feminist. The first answer is clearly correct. Linda can be (1) and not (2) but not vice versa. Yet more people chose the second.

"How come you can have situations of this kind? How is it possible?" Zadeh says. "And there I make one suggestion: that people sometimes normalize." As with *orange*, they convert an intersection like *pet fish* into a full-blown set, so its highest membership is 1 rather than 0.5 or 0.3 or 0.2. Thus a guppy could have low membership in *pets and fish*, yet high membership in *pet fish*. Compare a guppy to other pets and other fish, and it scores low. Compare it only to other pet fish, and it leaps up. The field narrows and the best examples change. The scale is different.

Or imagine an item which is part table and part cup, perhaps a toy table with a broad hollow on top. This ungainly goblet is not much of a table or a cup, so it is a *table and cup* to very low extent, say, 0.02. But it is a *table-cup* to high degree, perhaps 0.95. The mind forms a prototype of table-cup and compares the item to that, not to tables and cups. Osherson and Smith, Zadeh notes, compare to tables and cups, cite the difficulties, and wave all fuzzy theory away.

In the Linda experiment, the secret of the correct answer lies in seeing that *teller-feminist* is a subset of *teller*. But erring subjects in effect treated *teller-feminist* as an independent set. They normalized. They envisioned a prototype teller-feminist and found that Linda resembled her more than a prototype teller. These two prototype sets need not have a set–subset relation at all. If Rosch is right and people view classes as their prototypes, this result flows quite easily.

Of course, Kahneman and Tversky had asked about probability, not resemblances or fuzzy sets. Their results thus also suggest that human beings naturally think in degrees of truth rather than likelihoods, and can have trouble with even simple probabilities.

Zadeh cites another upshot of modifying nouns with adjectives. Sometimes simply combining two words changes their import. For instance, *red hair* alters the meaning of *red*. *Ice cube* changes *cube*. We bring knowledge to these terms. In the past, *red hair* has always referred to a special kind of red, quite different from the red of the spectrum. Our experience perfuses language. In fact, if someone pours water in a glass and says it's *full*, we'd actually be surprised if it were 1.0 full. Experience tells us that in this case, *full* means around 0.9 full.

Obviously, fuzzy logic does not completely describe words. They remain too rich and mysterious, too freighted with implications. "Language, eight-armed, problematic, demiurgic, infinitely entrailed, must be honored," wrote philosopher Galen Strawson in 1952. "Its riddling, jokey, mischievous, metaphoric, flawed, lapsible, parapraxic life must not be repressed, but tolerated, pleasured, submitted to, enjoyed, and so revealed for what it is."[17]

The real problem lies not in fuzziness, but in modeling human thought, a task so towering that no one has even come close to it. Fuzzy logic at least partly describes words, and provides an entrée into the conjunction fallacy and other mysteries of the mind. Classical logic stands agape.

Zadeh at one point said he felt that Osherson and Smith greatly impeded the advance of fuzzy theory into psychology. Gregg Oden disagrees. "I think that's an exaggeration," he says. "That paper got a lot of attention, but I think not all that many people were impressed by the arguments in it. It'll get cited by people who don't like fuzzy and never did."

Howard Gardner notes an intriguing irony to this intellectual adventure. After thousands of years, computers finally gave us the perfect devices for emulating the classic theory of concepts. Yet just at that moment, the theory fell apart. Computers forced us to take crisp boundaries seriously, and when we went searching for them, we found they didn't exist and never had.[18] The greatest exponents of crispness—Socrates, Locke, the early Wittgenstein—had labored under the deepest illusion.

5

FUZZY ENGINE, FUZZY INFERNO

■

No theory is good unless one
uses it to go beyond.

—ANDRÉ GIDE

Fuzzy logic is saving us thousands if not hundreds of thousands of dollars on a yearly
basis.

—PETER HOLMBLAD

DNA, the brain, bureaucracies, and computers all share a feature in common. It is a deep, powerful, elusive property, the heart of information technology, the nub of culture, and the secret of life itself. It is control.

Control is manipulating the environment toward some goal. It is not necessarily utter dominion and can even be gentle nudges. Yet it is the difference between a paramecium and a pearl. The lowly paramecium can sense information from outside and take action. A morsel of food drifts by and the creature moves toward it. The paramecium affects its world as the pearl cannot.

The etymology reflects the gist. *Control* derives from the medieval Latin *contrarotulare,* which meant to compare "against the rolls," the cylindrical records of ancient times. Bureaucrats would receive information about a land parcel, match it with these rolls, and decide how much to tax or whether to arrest. In a similar way, the human body has countless control systems. For instance, the hypothalamus, a small organ in the brain, constantly assays the bloodstream for the chemical angiotensin II. When it reaches a certain level, the "rolls" within the hypothalamus tell it to trigger thirst, and the higher the level, the keener the thirst.

All control involves information processing. An animal, company, or government must be able to acquire information (input), apply rules to it (process it), and act on the results (output). For instance, when a paramecium senses food, it matches that sensation against its inner records and moves toward it. To keep its democracy representative, the United States conducts a census every 10 years, applies rules, and reallocates the number of congressional districts. Most people think of control as just the power to act, but without good rules this power goes berserk, and without initial information it is oblivious.

Computers are machines that process information. Keyboards or other sensors take in information, software applies rules to it, and the upshot appears on video screens or in mechanical actions. Hence computers won some of their first triumphs in control, with the servomechanisms of Norbert Wiener, and have come to regulate a vast array of processes. Fuzziness entered the practical world with control as well.

THE IF-THEN MESH

When Zadeh wrote his first paper, he believed fuzzy logic would find its home in psychology, philosophy, and the human-oriented sciences, but as early as 1965 he was suggesting it would "play an important role" in control.[1]

In 1973, he published his second most influential article, which he calls "the key paper outlining a new approach to analysis of complex systems."[2] This paper laid the groundwork for fuzzy control. Indeed, it brought fuzzy logic partway out of the web of theoretical esoterica and into the daylight. It showed how engineers and corporations could use fuzzy logic.

In machines, control involves reaching a target and staying there. For instance, in thermostats the target is whatever number the dial points to. Suppose the temperature in the room is 70°F. Flip the thermostat down to 65°F, and the control system has a new target. It will take action to reach it.

How? In classic control, it uses a formula. In fact, the most common devices are called PID controllers, for proportional–integral–derivative, a mouthful whose three parts each refer to a mathematical operation.

In control, the road of the formula may be direct, but it may also lead up into the hills, winding about, forking left and right, before finally coming back down near where it started. Indeed, in complex cases, it may trail off into brush and yield no results at all. Yet often there is a shortcut that goes quickly from start to finish: human judgment.

Zadeh thought to capture judgment by fragmenting formulas into a string of fuzzy IF–THENs, such as

IF *temperature* is a little low, THEN increase it slightly.
IF *temperature* is moderately low, THEN increase it greatly.
IF *temperature* is very low, THEN increase it very greatly.

Such rules could describe a relation such as temperature to heat change just as fully as a formula, if more loosely.

Graphs show this approach more clearly. On a graph, a formula creates a line. Each value on one scale matches one or more values on the other. The result is a precise string of points representing all possible cases. But fuzzy logic, as Bart Kosko notes, creates overlapping fuzzy patches along this line. The smaller and crisper the patches, the more they resemble the line. When they become infinitely small and numerous, they are the line.

Why use this approach? If the equation is simple—like $5x=y$—there usually is no reason. It may do perfectly well and fuzzy logic could worsen the results. However, when formulas in control grow intricate and unwieldy, engineers can divide and conquer with fuzzy patches—with, in a sense, many simple equations (see Figure 6). Fuzzy patches can also be easier to derive and faster to use. Moreover, when control situations defy crisp formulas, engineers simply do not know the exact location of the line. Yet machine operators may know its approximate course, as shown by the vague rules of thumb they use to guide the machine. If so, fuzzy IF–THEN patches can describe it.

For more sophistication, Zadeh suggested a second approach: fuzzy algorithms. An algorithm is a step-by-step method of reaching a result.

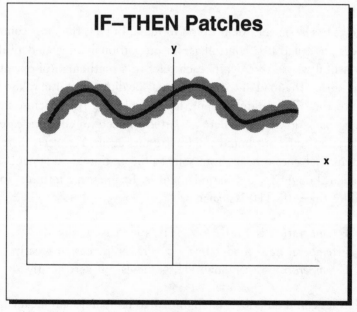

FIGURE 6

A recipe is a classic algorithm. We follow an algorithm every time we set up a tent, tie a knot, or drive a car. For instance, to use an automatic teller, we: (1) insert the card, (2) punch in a secret ID number, and (3) punch in specific commands. Among programmers, an algorithm is the procedural backbone of software. First it does A, then it checks B, and if B is greater than 100 it does C, and so forth. Algorithms are critical because they set forth the links in the chain of solution. If the algorithm is flawed, it doesn't matter how tight and bug-free the final coding is. It won't work.

Fuzzy algorithms are ubiquitous in everyday life. The steps in driving to the beach or tying a shoe are not exact, but fuzzy. Even recipes, which seem to require precise quantities, normally succeed with approximate ones. Indeed, it is the very imprecision of fuzzy algorithms that makes them practical. A highly crisp algorithm, such as that for walking a tightrope, is much harder to follow. Zadeh mapped out four kinds of fuzzy algorithm:

- Definitional, which categorize fuzzy input.
- Generational, which create fuzzy output.
- Relational, which describe systems in flux.
- Decisional, which issue commands from ongoing feedback.

Definitional algorithms funnel individuals into Rosch-like classes. Imagine a computer trying to recognize the letter A. An A is really a class and includes a welter of specimens in odd fonts and handwriting, most of which human beings decipher at once. But crisp algorithms treat A as almost a Platonic Ideal, prescribing the exact length, position, and angles of its lines. An A on a page must match this Ideal near-perfectly for a machine to identify it. Hence the crisp algorithm is not very useful. In contrast, fuzzy definitional algorithms give computers peripheral vision. They recognize the fringe members as well as the prototypes.

Generational algorithms create rather than define. Zadeh noted that they might churn out handwritten characters, cooked food, music, natural language, and speech. He described in detail examples of writing a capital P and baking chocolate fudge. A fuzzy algorithm could make good fudge because, again, the steps aren't precise.

Relational algorithms describe. If a person pumping up a balloon wished to know how much more air she could add before it burst, she could use a relational algorithm, one that sets forth current state, act, and result. A relational algorithm shows what is happening and would happen. With the balloon, the three concerns are the air inside the balloon (current state), the amount it increases (act), and the surface tension (result). A fuzzy relational algorithm might express the system as follows:

1. IF the amount of air is *small* and it is increased *slightly*, THEN the surface tension will increase *slightly*.
2. IF the amount of air is *small* and it is increased *substantially*, THEN the surface tension will increase *substantially*.
3. IF the amount of air is *large* and it is increased *slightly*, THEN the surface tension will increase *moderately*.
4. IF the amount of air is *large* and it is increased *substantially*, THEN the surface tension will increase *very substantially*.

Finally, decisional algorithms weigh feedback and apply rules of strategy or decision. Common decisional algorithms we use all the time, almost unthinkingly, include those for moving an object, going through a checkout line, and driving a car. Zadeh offered a simple algorithm for crossing an intersection with a stop sign:

1. IF no stop sign on your side THEN IF no cars in the intersection THEN cross at *normal* speed ELSE wait for cars to leave the intersection and then cross.
2. IF not *close* to intersection THEN continue approaching at normal speed; go to 1.
3. *Slow down*.
4. IF in a *great* hurry and no police cars in sight and no cars in the intersection or its *vicinity* THEN cross the intersection at *slow* speed.
5. IF *very close* to intersection THEN stop; go to 7.
6. Continue *approaching* at *very slow* speed; go to 5.
7. IF no cars *approaching* or in the intersection THEN cross.
8. Wait a *few* seconds; go to 7.

What use is such a procedure? Robots can use it to navigate a living room without bumping into chairs and walls. It also shows clearly how fuzzy algorithms can embody human know-how and experience.

This article appeared in January 1973, and later that year Zadeh spent six months as a visiting scientist at IBM. Unlike his earlier stay, he says, "I went there with the understanding I'd work on fuzzy sets and there'd be some people at IBM who wouldn't mind working with me on fuzzy sets." It never happened, and only a sprinkling of engineers attended a talk he gave on fuzzy sets. "Let's put it this way: at no point when I was at IBM did IBM take any interest in fuzzy set theory." So he joined a group headed by Ted Codd, who invented the concept of relational database. "I didn't try really to push my ideas," Zadeh adds. "I tried to learn, rather than propagate my ideas."

Had IBM seized upon fuzzy IF–THEN rules and developed them, the world might have seen fuzzy appliances a long time ago. However, Zadeh believes the opportunity was slightly premature. "If at that time, I could have shown them a fuzzy consumer product, IBM probably would have taken interest," he says. Yet he also notes that IBM had for years fallen prey to the NIH ("not invented here") syndrome. "It viewed itself as the vanguard of computing," he says, "so that anything done outside was poohpoohed. And eventually IBM suffered for that."

But IBM also lacked a specific, pressing problem to solve. At that very moment, one had arisen across the Atlantic, where fuzzy IF–THEN rules were about to spark the first fuzzy system.

THE STEAM ENGINE THAT LEARNED TOO MUCH

In 1973, professor Ebrahim Mamdani ("mom-DAH-nee") of London University was supervising a Ph.D. student named Sedrak Assilian. Assilian had built an engine with a little boiler and piston, and was seeking a way to control it automatically, to keep the boiler pressure and piston speed constant.

Assilian wanted the device to learn from himself. He regulated the pressure and speed. Every now and then the computer would measure the pressure and speed, he would adjust it, and this information was supposed to enlighten the system. But the machine was too eager and indiscriminate. It was learning everything, his mistakes as well as his improvements. He brought the problem to his advisor.

Ebrahim Mamdani was born of Indian parents in Tanga, a town of wide streets and palm trees on the Tanzanian coast near Zanzibar, with communities of blacks, Indians, and whites (mostly Greeks, Italians, and Britons). His father ran a small shop and other relatives worked as clerks and accountants for sisal companies. He lived there until he was 17 and remembers it as "fantastic," with cricket and volleyball matches, balmy days full of sun. He always felt relaxed. He read widely and recalls the 1956 Suez Incident vividly. "In many ways, kids at my school were more informed about things outside Tanga than many kids in Europe and America are today," he says.

He grew up polyglot, speaking Gujarati and a dialect of it, Kutch, from India; English, from school; and Swahili, from his African servants. "I now work with Germans and French and I don't know any of their languages," he says. "All my linguistic energies went into the wrong languages."

An excellent student, he faced a problem after high school. Colonial Tanganyika had no universities. "You could go to England, but that was very expensive," he says. "So the next best thing was to go to India and find out about your mother country. That wasn't easy either. There was no family left in India any more."

He went to Pune (Poona), near Bombay, and did his undergraduate work in the College of Engineering. Then, realizing he was "overeducated for Africa" and that its future "was a little doubtful because independence had come," he went to Britain for his Ph.D. He later became a professor of electrical engineering at Queen Mary College, London University.

When Mamdani saw the steam engine that learned too much, he puzzled over it and recalled Zadeh's recent article on fuzzy IF–THEN rules. "So we said, 'Why don't we try that?' " It took just one weekend. "I remember it well," he says. "My home was then 15 minutes from the college and I went in on a weekend and we did it. We were very impressed by how easy it was."

As Brian Gaines says, "They immediately got a working controller for the steam engine. That in itself was astonishing. And what they got was better than what they could have gotten from numerical methods and modeling. That was astonishing in its own right."

"It was really the turning point," Zadeh says. "It showed these concepts could be applied."

Their little steam engine is the grandfather of today's fuzzy systems— the washers, camcorders, and subways—and it repays a little probing. It is also not obvious from the theory, and scientists who read only the theoretical journals can mistake its nature.

The apparatus was highly simplified. "It was a toy," Mamdani says. "The whole contraption stood on a tiny wooden splint no more than one foot by two."

It had two parts: boiler and engine.[3] Each sought a target. The boiler's was a specific pressure—whatever Mamdani or Assilian set—and it reached that goal by adjusting the heat. Raising heat increased the pressure while cutting heat diminished it. The engine's target was the right piston speed, which it attained with a throttle valve. Opening up the valve boosted the speed and closing it down decreased it.

The steam plant had sensors that constantly monitored the boiler, showing its current pressure.* The device of course also knew the target. If current pressure matched the goal, all was well and the controller simply oversaw the situation, like a night watchman strolling down a street. If they differed, the controller had to act.

Its task was tricky. The relation between pressure and heat was highly nonlinear, meaning it did not change evenly. In a sense, different rules applied at different intensities. With linear equations like $5x=y$, adding two equations yields a valid third equation. The relationship remains the same with large amounts or small. Nonlinear equations defy this simple cumulation. They change as quantities increase or decrease, and can become mathematically intractable. Yet they describe the most important phenomena of life.

Mamdani and Assilian did not try to apply a formula to this problem, but rather broke it down into fuzzy IF–THEN rules. They began by setting up the fundamentals. Because the system used information of three kinds, they created three sliding scales. Two showed input, that is, information about the state of the boiler. The third was output, a command.

The first scale registered

> ■ *Pressure error*. How far off is the current pressure? Pressure error was the difference between current and target pressure. The fuzzy controller cared nothing for the absolute pressure. Only the distance from the goal, positive or negative, mattered.

---

* This description addresses only the boiler, since the engine worked exactly the same way.

The second scale measured

■ *Change in pressure error.* How rapidly is the pressure nearing or receding from the target? This information was crucial. Suppose a pedestrian is about to jaywalk and sees a car coming down the street. In deciding whether to cross, he weighs not just the car's distance, but how fast that distance is changing—its speed. If it is near but crawling, he may get across easily. If it is farther away but roaring, he may not. Error shows the present, change rate the future. In the engine, this input showed the difference between the pressure error an instant ago and the current error. Positive change indicated approach to the goal, negative change retreat.

Finally, the controller had to issue a fuzzy command based on these values, so a third fuzzy scale arose:

■ *Heat change.* What is the right response? Heat change described how much to adjust the heat.

Once the three continuums were in place, Mamdani and Assilian placed a cordillera of seven triangles on each one, all hedges:

1. PB—positive big
2. PM—positive medium
3. PS—positive small
4. ZE—nil
5. NS—negative small
6. NM—negative medium
7. NB—negative big

These overlapping fuzzy sets all had the same size and shape, and were spaced evenly along each line, but if control were more critical in the target zone, Mamdani and Assilian could have placed narrow spires near zero and flattish gables farther away from it. In any case, these formed the basic units of the system.

Next, the pair had to build the network of rules using these sets.
Assilian's original device began in a state of ignorance, a tabula rasa. He therefore had to train it. The new fuzzy device commenced in a state of knowledge. "We said, 'Instead of the computer learning from you, you'd tell it what to do with these rules,'" Mamdani says. They provided it with 24 rules, all in a PDP-8 computer.

Pressure Error

Change in Pressure Error	Negative Big NB	Negative Medium NM	Negative Small NS	Nil ZE	Positive Small PS	Positive Medium PM	Positive Big PB
NB				PB			
NM				PM			
NS				PS	NS Rule 3		
ZE	PB	PM	PS	ZE Rule 2	NS Rule 1	NM	NB
PS			PS	NS			
PM				NM			
PB				NB			

FIGURE 7

Mamdani and Assilian found these rules the very soul of the process. They not only described the workings of the engine in simple terms, but captured the experience of skilled operators, who express their expertise similarly. The rules enabled the steam engine to respond to verbal instructions rather than numerical ones.

Moreover, they worked in a special way—like a committee. The readings from the sensors fired several rules at once to different degrees. Each firing was like a recommendation from a committee member well-informed on a facet of the whole. Like a sensible chairperson, the controller assessed the recommendations and merged them into a crisp command.

The process may look complex, but in fact nontechnical people can easily understand it, a major virtue. For instance, in Figure 7, the fuzzy sets for pressure error lie across the top and for pressure change down the left. When combined, as in *if pressure error is positive small (PS) and change rate is nil (ZE)*, they yield a heat change command like negative small (NS) in the center.

Many of the squares are blank. Why? They are not necessary. One could place rules here, but in practice the system will not call on them.

Each box is a rule, a committee member. For instance, one is the expert on the case where pressure error is slightly positive and is not changing much. The command here is to reduce heat negative small, and this member recommends the right degree of negative small.

The three darkened rules in the matrix are the committee members offering opinions in this case. They are

> *Rule 1.* IF pressure error is a little positive AND it is not changing much, THEN reduce heat a little. (IF PS AND ZE, THEN NS.)
> *Rule 2.* IF pressure error is around zero AND it is not changing much, THEN don't change heat. (IF ZE AND ZE, THEN ZE.)
> *Rule 3.* IF pressure error is a little positive AND it is getting a little worse, THEN reduce heat a little. (IF PS AND NS, THEN NS.)

A true fuzzy controller would merge more rules, but it does not matter for this example.

Figure 8 shows how the recommendations arise from the rules. Thick postlike uprights stand at the zero points, where there is no pressure error or change, and triangles indicate the fuzzy sets. The two vertical lines that slice through all three rules are the sensor readings for pressure error and pressure change. Where each line cuts a triangle, it yields a value, such as 0.8, that is its membership in the set

Firing Fuzzy Rules

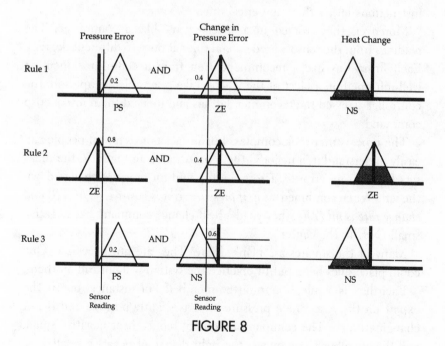

FIGURE 8

All three rules fire at once. In the first, the vertical lines yield a membership of 0.2 in positive small pressure error and 0.4 in zero change. The rule then takes the lesser of 0.2 and 0.4—that is, 0.2—and applies it to the negative small heat change command. In other words, it recommends negative small heat change to 0.2 degree.

Likewise, the second and third rules select the smaller of the two memberships and recommend the heat change command to that extent. At the end of the first stage, then, the controller has an assortment of degrees to which the heat change command should fire. They are negative small, 0.2; zero, 0.4; and negative small, 0.2.

Next, the controller must total up the recommendations.

In the most common approach today, shown in Figure 9, the controller lops off command triangles at the recommendation level. It then aligns the posts, superimposing the trapezoids atop one another to create a new geometric shape, like the shaded area at the bottom of the figure. This shape is the sum of the recommendations.

Bart Kosko has sharply criticized this superimposing. It junks information, he says, since it lets some trapezoids block out others. For instance, in this example, the recommendation from Rule 1 lies completely behind that from Rule 3. One committee member might just as well have remained silent. Such losses do not matter in simple systems. But in complicated ones, he says, they make all output tend toward zero. More information leads to less discrimination, which is just what should not happen.

Kosko has developed another approach which adds the trapezoids together, so they pile atop each other like Ossa on Pelion. The upshot usually resembles a bell curve, peaking over the center of gravity. This method, he says, exposes the full area of all the sets, forfeits no information, and yields more accurate results.

Finally, the controller must decide. It must defuzzify, or resolve all the recommendations into a single crisp command. Defuzzifying may seem a treacherous art. As Bertrand Russell suggested, we can imagine a blurred photo of a person from a sharp one, but the reverse is much harder. Mamdani and Assilian now had to make the blurred photo sharp.

The task turned out to be simple enough, for in fact we commonly

Defuzzifying

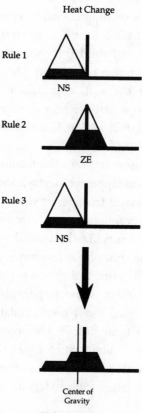

FIGURE 9

distill a range of numbers into a single crisp one. An average does exactly that. The mean and median perform similar feats. In all three cases, some information disappears, but a crisp figure emerges, and we may need it.

In fuzzy systems, the committee head is ultradiplomatic, always compromising as exactly as possible. The controller looks to the center of gravity of the geometric shape. It finds the point which here the shape would precisely balance on a straightedge, where a vertical line would slice it into two equal masses. This value is crisp and leads at once to a numerical command. In Figure 9, for instance, the center of gravity is just to the left of the zero post, so the controller would issue an instruction to reduce heat by that amount.

If this lone pass through the controller does not seem to accomplish enough, it is partly because a real controller addresses all the rules at once. Most apply to 0 degree, but even so there would be more rules than appear here. In addition, the controller does not just issue a command, then stop. It spews forth thousands of commands per second. Responding quickly to feedback, it can regulate changing conditions very closely and attentively.

Mamdani and Assilian's fuzzy controller exceeded a conventional device on several counts (see Figure 10). First, they found that error gradually diminished toward the goal, much as a ferry pilot nearing the docks aims more and more accurately until the boat pulls into the slip. Thus, the steam plant tended to come right in on target. In contrast, the digital controller usually overshot the goal. The system then had to compensate to bring it in.

The fuzzy controller was also swifter. Its speed stemmed partly from its parallel operations. By firing several rules together, it cut the time from here to there. However, it also exploited better focus. Instead of trotting out a complex equation describing every state to handle just one state, fuzzy control addressed only the states that mattered.

Performance of Controllers

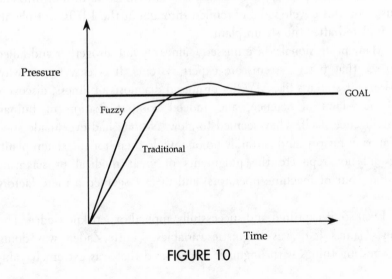

FIGURE 10

Moreover, fuzziness cut the risk of error. Conventional control uses one or few equations, so if one is wrong, it can skew almost any result. But fuzzy control compartmentalizes. It uses independent rules, so if one is wrong, it does not taint the whole. It is, again, like a committee, which can reach a reasonably good decision even if one member errs.

There were further advantages. Fuzzy control requires less information, up to 1/10th as much, according to Omron Corporation, and it can thus slash development time. Finally, since it works in linguistic terms, it can take information straight from people's mouths. It can absorb human knowledge without translation into a complex mathematical model.

The pair showed the device to Brian Gaines, then editor of the *International Journal of Man-Machine Studies*. "He said, 'You must write a paper for my journal,'" Mamdani recalls. "So I did." Mamdani credits Gaines with championing it to a wider audience. He induced Mamdani to present a paper on the steam plant at a conference on multivalued logic in Logan, Utah, where it won top prize.

Gaines himself came to doubt that fuzzy logic *as formal logic* was essential to the process. "What really happened was Assilian and Mamdani had invented expert systems and I couldn't say that. The word wasn't in use at the time." If Gaines is technically in error—DENDRAL had appeared in the late 1960s—he is correct in principle. The seminal expert system was MYCIN, which Edward Shortliffe and his associates developed and refined throughout the 1970s, but plainly completed after the steam plant.

Many professionals have missed Gaines's point altogether and failed to see that fuzzy systems are expert systems. It is easy to see why. AI expert systems like MYCIN promised diagnosis of illness, discovery of new laws of science, and indeed computerization of human intelligence itself. They seemed to glow. No one had ever made such claims in history, and certainly no one made them for the steam plant. It did not replicate the judgments of great medical or scientific minds, but of machine operators, and in fact seemed a mere factory tool.

Even so, it captured and successfully applied expert knowledge. The key, Gaines felt, was linguistic variables. "Lotfi Zadeh was doing interesting things with linguistic hedges and that was extremely valuable," he says. "Whether you call that fuzzy logic is a different matter. And if you count that as fuzzy logic, then it's valuable."

LAWS OF THE INFERNO

After Mamdani published his results in 1975, a spate of other applications appeared. In Delft, Holland, Professor Van Nauta Lemke and W. J. M. Kickert developed a fuzzy controller to regulate the temperature of a warm water plant. In Lyngby, Denmark, Professor P. M. Larsen and Jens-Jorgen Østergaard began work on a fuzzy heat exchanger. In 1975, D. A. Rutherford applied fuzziness to a sinter-making plant at the British Steel Corporation. The fuzzy controller outperformed human beings and slightly excelled the conventional device.

However, the first commercial fuzzy system, the big advance, came in Denmark.

In 1973 Lauritz Peter Holmblad, a Danish engineer, had left the Technical University in Copenhagen to work at F. L. Smidth & Company, which made cement. Cement is one of the little wonders of modern life, a powder which becomes a thick paste on moistening and hardens into artificial rock. Like paper and plastic, it aids everyone every day, yet most people are utterly ignorant of it.

It begins as common sediment—clay, limestone, and sand. Manufacturers mix these ingredients with iron ore in a slurry. They then dry it and break the limestone down into free lime (calcium oxide) and carbon dioxide. Finally, they send it through a kiln.

The interior of a cement kiln is a wild, hellish place, between 1,250 and 1,450°C (bronze melts at around 950°C, pure iron at 1,535°C). It is a huge slanting cylinder—the Smidth kiln is some 16 feet wide and 420 feet long—that rotates about once a minute. Slurry enters at the upper end and over the course of three hours slides down to the bottom. During this period, the raw materials combine to form the complex components of cement: di- and tricalcium silicate, tricalcium aluminate, and tetracalcium aluminoferrite. The end product is the small stones called *clinker*. Pulverize them and you have cement.

Events occur in this seething realm about which even the operators are unsure. "Conventional controls don't work because there isn't a good model of the process," says Gaines. "You've got a bunch of stuff going in and they all break up, so it's not like a chemical reactor, where you put a few chemicals in at a certain pressure and you get a detailed model. It's just a mess."

Conventional methods could regulate the blending of ingredients and other kiln processes well enough. However, they could not master clinker formation. A key factor was burning the material at the right temperature. If it burned too hot, the clinker was hard to grind and the kiln wasted energy. If it did not burn hot enough, quality suffered. Hence it was important to negotiate a balance. That required human operators, who adjusted the amount of fuel, the amount of material fed into the kiln, and the kiln rotation speed.

It is not the most exciting job on earth. Hours might pass where no control at all is necessary. The process is so slow that people found it hard to cope with, and at night some even dozed off. It also defied human workspans. An operator might try diligently to maximize conditions, but after an eight-hour shift, another operator came on duty and could spoil the whole effort.

F. L. Smidth had attempted to control the kiln with computers as early as the 1960s, but models of the process proved too complex and unwieldy. In late 1974 Peter Holmblad attended a conference of the International Federation of Automatic Control in Zurich. There he came across a paper written by Mamdani, not on fuzzy logic, but on self-organizing controllers. At the end of the article, Mamdani mentioned that self-organizing controllers were complicated and that he had recently stumbled on a better method: fuzzy logic.

Holmblad saw its appeal at once. It focused on the operator instead of the process. "We'd said to ourselves, 'You can take a guy with no education and in eight weeks train him to operate this kiln. So why is it so hard to make a math model for this?'" Holmblad recalls. "We saw fuzzy logic as a way to make a math model of the operator."

After the conference he looked up Zadeh's 1973 article, which stoked his interest further. "In there it was very obvious that you could process control rules," he says. When he discovered a training text with 27 IF–THEN rules for regulating kilns, he became very excited. These rules summed up the operator's knowledge in words. "I could see how fuzzy logic could handle such rules. So I said, 'Why don't we make a computer process by fuzzy logic and try to come to the same conclusion?' "

Holmblad had supervised the master's thesis of Jens-Jorgen Østergaard and soon drew him into the project. They both found that the cement industry had grown skeptical of computer control, since others had tried to replace people with linear controllers and failed. "Our first job consisted in convincing the cement industry that automatic control of a kiln was possible," Østergaard says.

Holmblad and Østergaard began with some simple experiments and theoretical investigations. They also carefully inspected the kiln itself. Their results were promising and Smidth told them to continue—with one proviso.

"We had a funny discussion over the term *fuzzy*," Østergaard recalls. "The company is highly respected and asked us to find another name." The two tried in vain, and eventually the firm acceded to the original term. "It was wise," he says. "It's easy to remember and everyone can pronounce it."

The project was daunting, a venture into an unlit realm, Østergaard says. In their first trial, they placed the rules from Holmblad's book in a fuzzy system and assessed the results. It was clear that some rule-based system would work. In June 1978, a fuzzy logic controller ran for six days in the cement kiln—the first successful test of fuzzy control on a full-scale industrial process. Fuzzy control slightly improved on the human operator and also cut fuel consumption.

"In the 1980s we came across something called *expert systems*, and found what we were doing was a type of an expert system," says Holmblad. Indeed, with this 1978 demonstration, he and Østergaard became the first engineers to run an industrial process online with an expert system. "Most others come up with advice to the operator, and he can decide whether or not to follow it," Holmblad says. "We closed the loop. The operator can look at it and stop it, but it's basically a closed loop."

In 1980, they installed a fuzzy controller permanently in the kiln and published a paper on it, which attracted attention in the fuzzy community all over the world.[4] The pair had developed the first commercial application of fuzziness.

In fact, the cement kiln was not really the first permanent application. In 1979, Holmblad and Østergaard had applied fuzzy control to a lime-reburning kiln at a paper mill in Sweden. "It's a small part of the paper mill," says Holmblad, "though from a control view, the problems are the same." Only six months later did they install a controller at the cement kiln. However, their paper described the cement kiln and drew the notice.

Though milestones, Østergaard today recalls, "These two were primitive." For instance, the controllers could only perform control based on verbal rules. But Holmblad and Østergaard found some calculation was necessary. So they developed FCL, Fuzzy Control Programming Language, which somewhat resembled FORTRAN. In 1982, they installed a system based on this language at a cement plant in tiny Durkee, on the Burnt River in eastern Oregon. They have continued to refine the controller and have since sold hundreds all over the world. Holmblad estimates that 10 percent of the world's cement kilns now use fuzzy logic, and that 50 percent could benefit from it.

Holmblad was unique. Among the countless corporate engineers in Western companies, only he perceived the money-saving value of fuzzy control and only he implemented it.

However, completely independently, another individual had found an utterly different use for fuzziness. It did not involve control or fuzzy IF–THEN rules, and in fact derived from the simple notion of fuzzy sets, but it was a manifest breakthrough.

A NIGHT IN MONTE CARLO

It was 11 P.M. on a July night in 1979 at the Winter Sports Palace in Monte Carlo. Luigi Villa had just won the world backgammon championship and now sat across the board from a new opponent, a robot named Gammonoid. A purse of $5,000 awaited the winner, and a TV camera carried the contest by closed circuit to a 200-seat hall.

Gammonoid had a computer terminal embedded in its chest, linked by satellite to an office at Carnegie-Mellon Institute in Pittsburgh. There sat professor Hans Berliner, with his BKG 9.8 backgammon program running on a PDP-10 computer. Every time Villa moved, a man in Monte Carlo tossed the dice, and Berliner fed the roll into the program and relayed back Gammonoid's move.

The viewing hall stood almost empty. No program had ever beaten a world champion at any game, and not many expected it to beat Villa. Even Berliner was pessimistic.

The auspices were certainly bad. Gammonoid had a prominent role in the opening ceremony in the Summer Sports Palace. When the lights dimmed and the band struck up the theme from *Star Wars*, Gammonoid was supposed to glide onstage and entertain the crowd. Instead, it snarled itself in the curtains and, to Berliner's dismay, thrashed about there for five minutes.

Berliner was born in 1929 in Berlin and came to the United States in 1937. He received his bachelor's degree in 1954 and then held a variety of jobs, including one at IBM from 1961 to 1969. But his passion was chess. For 15 years he was one of the top 12 players in the United States. In 1968 he became the world champion of correspondence chess, or chess by mail, and reigned until 1972.

"Many people deprecate games, on the grounds that they are for kids," he says. "But games have the essence of reality in them. They're miniature reality. If we can't understand how games work, we can't possibly understand how reality works. So, in a sense, games are a good place to test theories."

He was also interested in computers. His world championship naturally encouraged him, and, he says, "I thought to myself, 'Maybe I can make a contribution.' " He decided to return to graduate school and in 1975 received his Ph.D. from Carnegie-Mellon, where he now teaches.

He began writing chess programs. But of course chess is very complicated. "Initially you're happy to get the computer to do anything at all," he says. If there is a problem, "the best way to remedy it is to isolate it and work on it there." Hence, when he wished to investigate how to evaluate board position, he turned to a simpler game: backgammon.

Backgammon is also a game of skill, but more like poker than chess. Luck plays a much greater role. Players bet by doubling the previous wager, so bets increase geometrically: 2, 4, 8, 16, and so on. One either sees the bet or drops out and loses. Moreover, the best move depends on statistical calculations, and an experienced player can often sum up the situation on the board at a glance.

He labored on the backgammon program for three years. "I remember a really critical moment," he says. In the spring of 1978, he entered the software in an intramural university tournament. It lost its first two games to poor players and was eliminated. "I found it would completely outplay opponents, except it would make these really awful moves at critical points, at phase changes," he says. "It didn't understand."

Backgammon takes place in five to seven phases, and good players understand how to shift over. For instance, a player must know when to cease blocking the opponent and start running for home. The program was making catastrophic errors at these points.

"I started to sit down and think about what could possibly be wrong," he recalls.

He finally realized he had assumed that phase changes are crisp lines. They aren't. They are transition zones. For instance, at one moment it may be 45 percent important to block and 55 percent important to run for home. Hence, if the dice yield a much better blocking move than a running move, the player should block. Berliner's program would cross the Rubicon into running mode and not look back.

"I thought, 'Let's try fuzzy logic,' " he says.

He borrowed the idea of membership functions, but saw little else at hand. "As far as the rest of fuzzy logic, the implementations I saw: There weren't any implementations."

To his surprise, it took just a few weeks to recode. "The difference in play was incredible. I had the old program play the new one. The new one would win 70 percent of the time." Given the role of luck, he found this demonstration very convincing.

He then entered it in the world backgammon championship at Monte Carlo. "I didn't expect we'd win," he says. "I thought we had a moderate chance."

The program doubled Villa in the first game and beat him by taking a short-term risk for a long-term gain—exactly the approach of a good human player. Because of the double, BKG 9.8 then led 2–0. In the second game, the program doubled expertly and Villa declined, boosting the score to 3–0. The news began to spread, and more people filed into the viewing room.

In the third game, Villa built up an early lead and doubled. Though a human player might have withdrawn, BKG 9.8 remained in the game. Soon after, Villa's luck nosedived. BKG 9.8 redoubled and Villa declined. The score was now 5–0 and the television room was full of spectators, cheering and moaning as fortunes shifted. It resembled the

world championship itself, Berliner recalls, with fans arguing passionately about which moves were best and who had the better position.

Berliner himself was excited. "In my fondest dreams I might have
hoped for five or six points in a losing contest," he later wrote. "Yet here
we were, leading 5–0, with an excellent chance of winning the match."[5]
He had worried that an expert might be able to trick the program into
making disastrous moves, but it hadn't happened yet.

In the next game, Villa again built up an early lead and doubled. This
time the program declined, though spectators felt the decision was a
close one. The score was now 5–1, and the crowd awaited a Villa
resurgence.

In the fifth game, BKG 9.8 began aggressively and Villa took several
of its pieces. As a result, it established a defensive game. It then made
a stunning move that dramatically improved its position. Villa doubled
nonetheless and the program accepted. Soon after, it made another
excellent move that brought it out of defense and enabled it to start
running for home. Then its luck improved even further, and the
program brought all its pieces home before Villa. It had won the match,
7–1.

Berliner could scarcely believe this result. Spectators rushed into the
room, flashbulbs went off, reporters interviewed him, and experts
congratulated him. Villa, trounced by a machine so soon after his world
championship, was despondent, but Berliner graciously told him he was
the better player.

And, in fact, when he analyzed the games later, he concluded that
Villa had made the better moves technically. BKG 9.8 had committed
mistakes, none serious, but it had also made dazzling plays. It had done
well enough that, with a small advantage in luck, it prevailed.

In June 1980, he described the match in a cover article for *Scientific
American*. Unfortunately for the field, he failed to mention fuzzy logic
by name, but he attributes the program's final leap forward to the
membership functions. "When I wrote this up, the people in artificial
intelligence acted as if it couldn't possibly be right," he says. "The
people in fuzzy logic just applauded."

He then placed fuzzy logic in his chess program HITECH, since chess
also has fuzzy phase transitions. For instance, at the start, a player seeks
to develop the pieces. In the middle game, that concept vanishes. In the
end game, one wants to move pawns quickly down to the end. "All these

are on sliding scales," Berliner says. "The degree of sophistication is amazing. Some of these things have three or four variables as to what is important."

HITECH is now the second best in the world, with a rating of 2410. It is not just software, but a special-purpose machine, with thousands of chips and hundreds of thousands of decision-making devices. It is like a rack of electronic equipment with six boards that fit on spindles and swing back and forth.

Berliner says all the top chess programs use fuzziness now, though his remains a bit more sophisticated. "They all know there's no sharp division between phases."

FADEOUT

Chess software, however, was both an exception and a tangent. The true bounty for engineers and corporations lay in fuzzy control. As Holmblad and Østergaard installed the kiln and published their findings, one or two rival cement companies imitated them, but otherwise no American or European firm followed suit. In fact, interest in fuzzy logic began to wane, and the decline continued into the 1980s.

Ebrahim Mamdani found much resistance in Europe. "If you're looking for an adverse reaction, it was from the control engineering community in the U.K.," he says. He felt he had discovered a promising technique, so he applied for a grant from the Science Research Council, Britain's equivalent of the NSF. "We simply asked to be allowed to do more work on this, to find out why it works so well, and what its characteristics were. We asked for about £10,000 to £15,000, which wasn't very much. They were not interested. They said it wasn't timely." Mamdani speaks of this setback with disappointment, though he adds, "I think it's understandable. It was early in the paradigm."

He did obtain funds to investigate opinion among U.K. cement manufacturers on the controller. At first, the cement industry was not interested. "I rang up one cement company and said, 'You should see this. They've used fuzzy logic.' They said, 'No, our problem is not controlling the kiln. It's how to clear away the clinker that sticks to the side of the kiln.' "

After the early success at Smidth, the cement industry changed its mind. Companies approached Mamdani, and he received several small grants from the Science Research Council. But eventually interest withered. "In the cement industry, everyone wanted to do it by themselves," he says. "Gradually it all petered out. The funding wasn't adequate. There was tremendous antagonism toward fuzzy logic throughout Europe."

Why? Mamdani has two hunches.

First, fuzzy logic was actually too easy. It therefore failed to engage the Western research ethos. "It has to have sufficient intellectual merit, which is normally measured in how much mathematics there is. And fuzzy, which is practical, is considered by them too simple, not intellectual enough." Mamdani recalls one Japanese from a famous lab who returned from a robotics conference at MIT scratching his head. He didn't understand half of it. There was too much math and theory, he said, and not enough engineering.

Second, fuzziness simply went against the grain. Mainstream scientists could not accept it, and they ultimately controlled the funding. "For any original worker, you cannot work against your peer group," he says. "If your peer group is interested in a particular area, they determine what is worth working in."

It was an epitaph for fuzziness. "As far as the Western world, it just died," says Ron Yager.

In fact, engineer Richard Tong, who had worked with Mamdani earlier, issued a code blue on fuzzy control around 1980. He pointed out that fewer papers were appearing. "I was suggesting that was an indication people were no longer making any interesting progress and perhaps we had to rethink what the technology was all about," he says. "Particularly in the States, there's been a significant amount of resistance to the technology and it was hard to get any money to continue the work, and it basically died out here until the last few years." The funding lay in AI and traditional expert systems, and hence Tong ceased work on fuzzy logic.

So did Mamdani. "I had to make a conscious decision not to encourage students to go into the field," he says, though engineers from China and Japan wanted to come to his lab. He turned his attention away from fuzzy logic, to applying knowledge-based systems to control.

It even affected Zadeh, the Ayers Rock of confidence. "Lotfi has always said he has thick skin," says Yager. "I think he was optimistic all along, then around the early 80s, I got the feeling that it was wearing on him. There was a certain tiredness, I guess. Lotfi would never admit this."

But Zadeh does concede that the field reached a nadir in the early 1980s. Though he wrote much on fuzzy expert systems and natural language, he too more or less dropped fuzzy control. At UC Berkeley, he taught database classes and Introduction to Information Processing. "I wouldn't talk about fuzzy logic in those courses," he says.

The irony was considerable. Just as fuzzy logic yielded its first great prize, its audience slid away. However, the discoveries of Mamdani and Holmblad were not quite trees falling in an empty waste. A few people heard them on the other side of the world, where they resonated with some of the oldest ideas in history.

6

THE VAGUE ARCHIPELAGO

■

Oh, East is East, and West is
West, and never the twain shall
meet.

—RUDYARD KIPLING

The West lives in a world separated into two terms: subject and
object, self and non-self, yes and
no, good and evil, right and
wrong, true and false. It is therefore more logical or scientific,
where yes cannot be no and no
cannot be yes.

—SHŌSON MIYAMOTO

■

It is a well-known optical illusion. Place a vertical line near a horizontal line of equal length and ask which is longer. Americans tend to pick the vertical, but Suku tribesmen in Zaire overwhelmingly choose the vertical. Why the difference? The Suku live in grasslands where the horizon is aways visible, and they are always interpreting slight rises in the distance. They focus more on verticality, and it makes a deeper impression on them.

The lesson is somewhat unnerving. Culture and experience affect our judgment at automatic levels. We believe our assessments rational and objective, yet we cannot escape our habits of thought. We live with ingrained and invisible biases.

Fuzziness found a ready welcome in the Far East. Japan now has over 1,000 fuzzy specialists, and Wang Peizhuang estimates China has some 10,000, most developing special-purpose industrial applications little known in the West. In the United States, by contrast, fuzzy professionals number in the hundreds.

Why did the Japanese seize upon fuzzy logic? The question has no simple answer, but like the Suku, the Japanese focus on a special feature of their environment: vagueness. It is more central to their lives, and they notice it more quickly and handle it more adroitly. In addition, they are highly pragmatic, exploiting useful notions regardless of the broader theory. They also have an extraterritorial curiosity, a penchant for looking over the fence, and it has taken them all over the globe to their fabulous profit.

THE GREAT SQUARE HAS NO CORNERS

Eastern philosophy overflows with concepts rarely expressed in the West, except by a few unconventional thinkers like Heraclitus and Hegel. Part of this intellectual heritage arose in India and much else in the long history of China itself.

India's rich tradition of logic is almost unknown in the United States and Europe. Indian logicians have defended almost every conceivable position on the Laws of Contradiction and the Excluded Middle, and in their heyday, around the 6th century A.D., most favored them. In fact, some schools took the Excluded Middle to its extreme, maintaining that the world contained either an item or its opposite. Stages in between did not exist. Indeed, Hinduism still holds that the physical world is *maya*, illusion.

However, there were at least two influential exceptions: the Jains and the Buddhists.

Jainism is an ancient religion, much older than Buddhism, and some Jains date their creed as far back as Mohenjo-Daro (c. 2500 B.C.). The Jains stress tolerance. Indeed, they take the doctrine of *ahimsā* (nonviolence) to extremes. They abjure agriculture for fear a plow might maim a bug, and some sweep the ground before them when walking to prevent similar mishaps. Their ascetic philosophy greatly influenced Gandhi, who adopted *ahimsā* almost intact.

The Jains developed a distinctive logic, best illustrated by their parable of the blind men and the elephant. A group of blind men encounter an elephant and explore it by hand. One feels the ear and declares it is a fan. Another touches the leg and thinks it a pillar. A third pats the body and says it is a wall. A fourth grasps the tail and states it is a thick rope. All are partly right; none is even close to wholly right. Yet through synthesis, a larger truth emerges.

The world is this elephant, say the Jains, and each of us knows only the small part we contact directly. They call this concept *syādvāda*, and according to one scholar, it "emphasizes that truth is many-sided. . . . Reality is complex like the many colored dome."[1] On this foundation, the Jains built a formal logic in which existence and nonexistence inhere in everything, and thus every statement is partly true and partly false. There is no certainty, or at least none we can know.

Jaina logic reached Japan, if indirectly. The Ryōanji rock garden in Kyoto, perhaps the most famous in the world, has 15 stones on a field of raked white gravel. Yet no observer ever sees all 15 at once. No one sees everything.

More influential, though more elliptical, was Sākyamuni (563–483 B.C.), better known today as the Buddha or "Enlightened One." He lived in India during a time of turmoil, in which larger states were coalescing through bloodshed, treachery, and the quashing of tribal institutions.

These upheavals stirred him to emphasize nonviolence and exalt wisdom over wordly goods and power. Indeed, Buddhism resembles Jainism in many ways, and some scholars suggest they have a common base.

Buddhism has distinctly fuzzy elements, to the point where Bart Kosko can say, "My claim is the Buddha was really the world's first fuzzy theorist." For instance, the Buddhist notion of *prajñā*, or wisdom, refers generally to intuition, a much deeper, broader, more mysterious entity than reason. It extends beyond the infinity of space and time. It demands immediacy, without reflection of any kind. It is one, dividing nothing into two, even uniting see-er and seen, actor and acted upon.

These fusions turn the Law of Contradiction completely inside out. For instance, one verse in the *Prajñāpāramitā Sutra* runs, "Because there is no Buddha, there is Buddha, and because there are no sentient beings, there are sentient beings." Whereas Buddhist reason, or *vijñāna*, endlessly analyzes, divides, separates, *prajñā* unites, makes the world whole. Thus, according to Suzuki, *prajñā* embraces paradox. *Prajñā* "has its own way of dealing with this world of dualities. The flower is red and not-red; the bridge flows and not the river; the wooden horse neighs; the stone maiden dances."[2] *Prajñā*, he says, "swings the staff; sometimes it asserts; sometimes it negates and declares 'A is not-A and therefore A is A.' This is the 'logic' of *prajñā*-intuition."[3] It is very far from the logic of Aristotle.

Buddhism crossed the Himalayas into China in the 3rd century A.D., where a variety of other fuzzy philosophies had long been brewing.

By tradition, Daoism springs from the *Dao De Jing*, the work of Lao Zi, whose name means "Old Master." We know nothing about this author, and at least one scholar believes the work is simply a medley of sayings from many sources. They arose in a period of oral composition lasting from about 650 to 350 B.C., and the first manuscript was written around 200 B.C. at the latest. It is a chocolate box of paradox which Joseph Needham called "the most profound and beautiful work in the Chinese language."[4]

It stresses the wisdom of seeing the contradictory in the whole. For instance, opposites on a continuum require each other:

Difficult and easy complete each other,
Long and short form each other,
High and low fulfill each other.[5]

Appearance belies reality:

> Great straightness seems crooked,
> Great cleverness seems clumsy,
> Great triumph seems awkward.[6]

In other paradoxes, states spawn their antitheses, or excess striving for a goal leads to its opposite. Some even emphasize a kind of imperfection in perfection:

> The great square has no corners.
> The great vessel is never completed.
> The great note sounds muted.
> The great image has no form.[7]

And on it goes, a throbbing tattoo: The superior person accepts apparent contradictions. Wisdom lies in paradox.

The duality of yin and yang expresses partial contradiction and has long pervaded Eastern thought. "No aspect of Chinese civilization—whether metaphysics, medicine, government, or art—has escaped its imprint," says scholar Wing-tsit Chan.[8] The terms originally referred to the shady and sunny sides of a mountain, and evolved to mean the two complementary powers of the universe. Yin is dark, complex, intuitive, contemplative, soft, receptive, feminine. Yang is bright, direct, rational, active, hard, creative, masculine. Yin is the sage, yang the ruler. The Chinese saw yin and yang in moon and sun, decay and growth, winter and summer. Imbalance of the two caused illness in the body and instability in the state. Indeed, the idea bred the profound Chinese sense of cycles in history, which has toppled emperors.

Yin/yang may sound like Aristotle run amok, but its key is fusion, rotation, yin becoming yang, yang becoming yin. Hence, it really represents harmony in contradiction and unity in multiplicity. For instance, its most famous emblem—the ancient *tai ji tu*, or "diagram of the supreme ultimate," shown in Figure 11—is a circle carved into two commas of light and dark, each with a dot of the other in its bulge. It suggests a spinning wheel, opposites whirling into each other. The dots symbolize the Daoist notion that once anything reaches its extreme, it

FIGURE 11

develops the seeds of its contrary. The *tai ji tu* adorns the South Korean flag, flanked by *Yijing* (*I Ching*) hexagrams.

Of course, the mere existence of a few books does not define a culture's attitude. If it did, one could claim Victorian England embraced contradiction on the basis of Oscar Wilde ("The only thing that ever consoles man for the stupid things he does is the praise he always gives himself for doing them"). But Buddhism and Daoism were religions, pervasive guides to life. They were not just available ideas. They saturated the mind.

As Chinese history unfolded, as new questions rose, they braided together with Confucianism and each other, evolving, untwining, and merging again. Around the 3rd century A.D., Daoism mixed with Confucianism to become Neo-Daoism, which in turn influenced early Chinese Buddhism. Daoism also affected Chan Buddhism (from the Sanskrit *dhyāna*, or meditation, later called Zen in Japan). Both Hua-yan (Flowery Splendor) Buddhism (7th century A.D.) and Daoism helped shape the Neo-Confucianism of Zhou Dunyi (1017–1073). And the great Zhu Xi (1130–1200), molder of orthodox Neo-Confucianism and the most influential philosopher in Chinese history, suffused his creed with partial contradiction, even as he tried to purge it of Buddhist and Daoist strains.

Unlike the Indians, the Chinese and Japanese made few contributions to logic. Indeed, China's most prominent school was that of the Dialecticians, and they were ultrasophists, parodists of logic. Perhaps the best known were Hui Shi (c. 380–c. 305 B.C.) and Gong-sun Long

(c. 320–c. 250 B.C.), the latter apparently author of the *Gong-sun Long Zi*, now corrupt. This work contains an assortment of paradoxes, some almost identical to those of Zeno the Eleatic.

Hui Shi once said, "The heavens are as low as the earth; the mountains are on the same level as the marshes."[9] Why? Clouds form below mountaintops ("earth") and marshes can lie high on mountains. Moreover, compared with the immensity of the universe, earthly features are all equally "low."

He also said, "Going to the state of Yue today, one arrives there yesterday."[10] Time is relative. Today is the yesterday of tomorrow, as yesterday is the yesterday of today.

Gong-sun Long offered the most famous paradox of all. According to legend, he was riding a white horse near the frontier when guards halted him, saying, "Horses are not allowed to pass." He replied, "My horse is white, and a white horse is not a horse." The flustered guards, the story goes, let him through.

Why is a white horse not a horse? First, *white horse* connotes qualities *horse* does not. Second, *white horse* excludes individuals that *horse* includes, such as gray horses. Third, the universal *white-horseness* differs from *horseness*. That is, Gong-sun Long interpets "A white horse is not a horse" to mean "The category *white horse* is not identical to the category *horse*."

These "paradoxes" go on and on, but the subtext is clear. Reason supports inanities. It can prove anything and therefore nothing. More generously, human thinking twists words and ideas, gives them varying senses, laughingly escapes the torpid rules. The practical Chinese tended to ignore the Dialecticians—"They can subdue other people's mouths, but cannot win their hearts," said Zhuang Zi (fl. 340 B.C.)[11]—and the upheaval of the times and competition from other schools sealed their demise.

Buddhism and Confucianism had sweeping impact in Japan. By the mid-1980s, over 83 million of the 120 million Japanese called themselves Buddhists. The Neo-Confucianism of Zhu Xi became the official ideology of the Tokugawa shōgunate (1615–1868), the incubator of modern Japan. Moreover, Shintō, the native religion of the islands, remains vigorous. It stresses the emotional, especially sincerity of feeling and action, which it can exalt beyond truth itself. With such a background, the Japanese not only tolerate vagueness, but as U.S. trade negotiators have learned, make an art of it.

Their language reflects it. In fact, it is not inherently vaguer than any other tongue, but the Japanese deploy it in an extraordinary way, hinting, hedging, adumbrating. "When I come back from Japan, I have a hard time," says analyst Sheridan Tatsuno. "People say, 'Stop beating around the bush.' When I go to Japan, I notice I have to be roundabout. Everything is subtlety and nuance. It's like San Francisco on a foggy day compared to a crystal clear day."

For instance, the Japanese rely heavily on context. They omit the subject and object of a sentence wherever possible. The same noun can be singular or plural, like *you* in English, and verbs provide no clue to their number. Verbs also lack the fairly precise indications of time found in Western languages.

The Japanese also make much greater use of ambiguities available in other languages. They insert numerous softeners like *chotto* ("a bit") into their sentences. They tend to blur sentences even when they know the exact answer. If asked, "How many yen do you have in your wallet?" a Japanese may say, "About 10,000," even if he knows he has exactly 10,000. Thus, when speaking English, Japanese say "I think" much more often than native speakers. Japanese sentences commonly trail off into a *but* or *however*, leaving the listener to infer the reservation. The term *maa* means, "Well, more or less—," and speakers employ it when they disagree with the listener but don't want to say so outright.

Of course, English can convey all these sentiments too. For instance, an American may say, "I think I have the wrong number" when she *knows* she has the wrong number. "I think" softens the utterance in a situation of potential offense. But the Japanese resort to these devices far more often.

Such habitual vagueness requires listeners to defuzzify more. And in fact, Japanese speech and literature prize implication, a refined indefiniteness. The brief, evocative poems called waka and haiku dominate the literature, and similar genres reigned in China for much of its history. Noh drama is also rich in suggestion. In art, rock gardens and landscapes evoke more than they state.

As many have remarked, Japanese society is notably conformist. The long communal tending of rice fields and the density of population has made them treasure social harmony. Conformist groups respond to implication rather than crisp command. They also tend to avoid face-offs. Hideki Yukawa, winner of the Nobel Prize for his research on the meson, says Westerners developed science because they confront nature. But the leitmotif of the Japanese mind, he says, has been a "tendency to sidestep as far as possible any kind of confrontation."[12]

The scholar Charles Moore offers perhaps the best summing up: "Precision, determinateness, direct logical analysis and/or exposition, and direct confrontation of any sort are simply out of order. Indirectness, suggestiveness, evasion or evasiveness, the smile rather than the logical argument, sentiment rather than logic and objectivity, a polite affirmative answer rather than frankness or challenging opposition, symbolism in many aspects of life and art and literature—these are all telling examples of the point."[13]

Japanese sensitivity to gray zones has technological consequences. For instance, scientist Kazuhiko Azumi points to robots.[14] They lie betwixt and between, neither human nor machine. While the West has a horror tradition of half-formed, violent humanoids—the murderous golem, Frankenstein's monster, the insurgent robots of R.U.R.—the Japanese make robots the charming heroes of comic books. Gray-zone creatures appealed to them, and gray-zone logic would too.

Vagueness also affects their attitude toward manufacturing. The Japanese tend not to think of products as finished or not-finished, but as partly finished. "For the Japanese, problems are rarely clear-cut, with simple answers," writes Tatsuno. "Problems are viewed as complex, incoherent, and vague, requiring incremental, diffused solutions rather than the hard-hitting, close-ended solutions commonly proposed in the West."[15] A TV or luxury car is an ongoing project, never quite done, and they improve it bit by bit until over the years it becomes outstanding.

One must not overestimate the role of vagueness. The Japanese round off constantly like everyone else, and they plainly grasp the value of precision. Some Japanese scholars oppose fuzziness as lustily as their Western counterparts. Moreover, when fuzzy products debuted in Japan, companies dubbed them *aimai*, Japanese for *fuzzy*. Consumers responded wanly. When the firms then changed the name to *faaji*, a meaningless transliteration of *fuzzy*, sales soared.

Yet the Japanese had no Aristotelian paradigm to overturn, and their sensitivity to vagueness would play a key role in Japan's acceptance of fuzziness.

THE PRAGMATIC SPUR

Japan is a necklace of islands flung out on the edge of Siberia, for most of its history far from the world's major trading nations—except China. It has almost half the population of the United States, yet less land than Montana, less farmland than Arkansas, and few natural resources of any kind.

The Japanese must thus live off their wits, and they have become intensely practical. They scorned the contemplative Hīnayāna Buddhism in favor of Mahāyāna Buddhism, a broader variety which held that one could find total truth in a secular life. Zen in Japan became more labor oriented, and, in an odd echo of Calvin, Zen master Shōsan Suzuki (1579–1655) told farmers, "Farming is nothing but the doings of the Buddha," and merchants, "Renounce desires and pursue profit single-heartedly."[16] The Japanese Confucian Ogyū Sorai (1666–1728) laughed at the meditative Confucians of the Song Dynasty: "As I look at them, even gambling appears superior to the quiet sitting and having reverential love in one's heart."[17]

Partly for this reason, the Japanese have generally shunned abstract speculation. They rarely spin out elaborate systems of ethics or metaphysics, and some branches of Buddhism, especially Zen, are fiercely hostile to abstract thought. As theoretical physicist Yukawa notes, "The Japanese mentality is, in most cases, unfit for abstract thinking and takes interest merely in tangible things. This is the origin of the Japanese excellence in technical art and the fine arts."[18] Indeed, with their focus on the concrete and their concern for space, the Japanese have excelled at the miniature—from bonsai to haiku. Some, like computer scientist Alan Kay, believe this trait along with their esthetic sense will make them outstanding microchip designers.

Japanese pragmatism received a boost from its two centuries of isolation under the Tokugawa shōgunate. Economically, it forced Japan to become self-sufficient—more than self-sufficient actually, since it had to feed a host of nobles, samurai, and other idlers. Behind the rice-paper curtain, capitalism flowered, free of debt to other nations. Much as gold and distance created an independent economic center in California, the long incubation of enterprise built one in Japan. The nation prospered, sea traffic thrived, and towns expanded. From 1600 to 1853, Tokyo grew from a mere village to a metropolis of over 1 million, the largest city in the world.

Most importantly, a money economy appeared. Coins replaced sacks of rice as the medium of exchange. The money economy, in turn, put the samurai in debt to the merchants, swelled the towns, blurred class distinctions, and ultimately undermined the feudal shōgunate and propelled Japan toward foreign markets.

Westerners tend to treat Commodore Matthew C. Perry's arrival in Tokyo Bay in 1853 as the lone event that sparked Japan to open its doors to Western thought and commerce, but in fact it simply capped a train of developments. Japan was ready to expand, economically as well as militarily. It had also seen China defy the West and suffer the grim Opium War (1839–1842). Japan therefore opened, but narrowly. It took exquisite pains to avoid the "spheres of influence" that were sapping China, and jealously guarded its markets. Foreign investors had proved calamitous to Turkey and Egypt, and it restricted them as well.

Its rise was swift. In 1894 Japan defeated China and seized Taiwan. In 1905 it humbled Russia in the Russo-Japanese War, astounding Westerners and winning dominance over Korea, which it annexed in 1910. In 1931 it took over Manchuria. In July 1937, Japanese and Chinese troops clashed briefly at the picturesque Marco Polo Bridge near Beijing, and Japan invaded China. This event, not Hitler's 1939 invasion of Poland, marks the true start of World War II.

By then Japan was a formidable economic power. It had built huge bank-centered conglomerates, the great *zaibatsu* like Mitsui and Mitsubishi. These were corporations on a scale hardly imaginable in the West. In the late 1930s, for instance, Mitsui employed over three million people and was the largest company in the world. It did business in banking, mining, manufacturing, and an array of other fields.

After the United States conquered Japan in World War II, the nation lay in ruins. Wounded ex-soldiers wandered the streets in rags, begging. Food was scarce, unemployment soared, and inflation and black markets ravaged the economy. At the same time, the occupying Americans imported U.S. movies, and Japanese sat in darkened halls dazzled by visions of refrigerators, cars, and highways.

As they slowly righted themselves, the Japanese evolved a strategy to seize some of that prosperity. It was a highly Japanese approach, but its central thrusts surprised Westerners. The Japanese badly needed capital and had plenty of cheap labor. By traditional thinking, they should have opened themselves to foreign investment and stressed labor-intensive industries. They did just the opposite.

First, they rebuilt the ancient walls of protectionism. They still disdain free trade, so Japanese consumers pay more for most Japanese products than Americans do. In effect, citizens underwrite the great companies. But the Japanese firms did not simply batten off the nation and grow fat and torpid. They used these subsidies to sell inexpensively abroad, and thus expand market share and widen the wealth funnel into Japan. Protectionism also insured a safe home market where Japanese companies could breed new industries without crushing competition from experienced foreign firms.

Second, they entered high-tech and adopted a coordinated technology policy. They would compete as a nation, while Western states vied as a motley of companies. They would harness their powerhouse firms in the same direction, while Western companies scattered and wasted their energy.

The job of steering course fell primarily to the Ministry of International Trade and Industry, or MITI. MITI carefully selected and rewarded promising industries and pruned off dead-enders. The Japanese penchant for consensus might have given MITI much influence in any case. But MITI could also offer direct financial aid, bank loans at nominal interest, and tax incentives. Moreover, the ministry attracted smart executives whose shrewd judgment further boosted its prestige.

Japan began its recovery with textiles and shipbuilding, but MITI quickly saw the special value of electronics. These products involved no costly raw materials, were generally lightweight and easy to ship, and seemed infinitely shrinkable. They mainly required competence, cleanliness, and attention to detail, all national virtues. Moreover, electronics was a rapidly expanding market, yielding more and more new products and the profits to go with them. The Japanese issued the world's first transistor radio in the 1950s. In 1957, Japan passed a law directing MITI to select worthy electronics projects, set production and cost goals, and fund them. As Clyde Prestowitz says, "The Japanese government had decided that the electronics industry was too important to be left only to businessmen."[19]

Everyone today knows the results of this plan. Indeed, the upsurge of Japan is one of the defining events of the late 20th century, rivaled only by the advent of the European Community and the demise of the Soviet Union. The Japanese today boast the world's biggest banks, securities firms, and electronics companies. They are the richest large nation per capita on earth, and they pursue one of the world's most advanced research agendas.

Technology has been a lead horse in their chariot, and the Japanese appreciate it. "I go there and watch television," says scientist Alan Huang of Bell Labs, "and when you turn on the television, there are always jokes about lasers and computers, and technology is viewed as great. Technology is the reason they have all this prosperity. Here in America, you see little twinges. Here it's basically a negative thing."

If the Japanese latched on to fuzzy logic faster than the Americans, one reason lies in their keen eye for the practical. Another lies in the fruits of that pragmatism. The Japanese today control the American market in consumer electronics. U.S. firms have not pioneered fuzzy logic in televisions because almost no U.S. firms make televisions.

THE GREAT SEARCH

In 1542, a Portuguese boat wrecked on Tanegashima, an island just south of Kyushu. It was carrying smooth-bore muskets, which amazed the gunless Japanese. The lord of Tanegashima bought two for a vast sum and gave them to his master swordsmith to duplicate. Technical problems stumped this artisan, but when a second Portuguese vessel appeared a few months later, he traded his daughter for gunmaking lessons and was soon making weapons as good as the originals. The technique quickly spread through the archipelago.

Japan's famous borrowing from other cultures is hardly new. Around 300 B.C. a prehistoric, nomadic society gave way to a strikingly different culture of rice-paddy farmers. Scholars once thought this break represented an influx of new people. They now believe that it simply shows fast assimilation of Chinese ideas, triggered by the spectre of a slowly unifying China.

When the Sui emperor Wen Di united China again in 589 A.D., the Japanese were also watching carefully. They sent four missions on the hazardous journey to Sui China between 600 and 614, and 15 to the Tang Dynasty from 630 to 838. Tang China was the largest, best run, and most sophisticated nation on earth, and the Japanese siphoned off major elements of its culture, including: Buddhism (from 624 to 692, the number of Buddhist monasteries in Japan leapt from 46 to 545), ideograms, the Confucian social structure, and methods of architecture, city planning, ceramics, and metallurgy.

The Japanese have excelled at such defensive adaptation. To grasp their sense of vulnerability, one must try to imagine the history of Britain if the Bourbons, Hapsburgs, and Romanovs had periodically united continental Europe for centuries. Westerners marvel that the Japanese have copied so extensively, yet remained themselves. Had they not copied so well, they might have lost their identity altogether.

However, Japan has not always adjusted smoothly. In 1549, Francis Xavier, founder of the Jesuits, landed in Kagoshima in southern Kyushu and quickly made converts. He had arrived during the blood-soaked Sengoku-Jidai, or Period of the Country at War (1534–1615), which crested on October 21, 1600, in the muddy Battle of Sekigahara. Through key desertions, Ieyasu Tokugawa emerged the victor. He established the Tokugawa shōgunate, and in 1614 banned Christianity. It smoldered underground, and in 1637 some 37,000 peasants erupted in the Shimabara Rebellion. They seized Hara Castle near Nagasaki, flew Christian banners, and held off government troops for three months, falling only when their food and supplies ran out. The shōgunate massacred virtually all 37,000, then halted trade and contact with the West. In 1640, Portuguese emissaries appeared at court to seek reversal of this policy and the shōgun beheaded them. Ultimately, the government allowed a trickle of goods to pass through the hands of Dutch traders almost imprisoned at Nagasaki. It stayed that way until 1853.

Why bolt the gates? The shōgunate had just united a warring nation and did not want foreigners supplying guns and rival creeds to distant regions unenthusiastic for its rule. But the government also forbade importing Western books, at a time when Galileo, Kepler, Newton, and other Europeans were remaking our view of the universe. A few brave Japanese endured imprisonment and death to learn Western technology. By the 18th century, the shōgunate was allowing some Western books into the country, but they were dazzling driblets. Hiraga Gennai (1728–1779), for instance, sold all his property to buy a Dutch book on natural history.[20] These works were in tongues scarcely comprehensible to the Japanese, and a few of them made painful, heroic efforts to decipher them word by word. By 1811, Japan had established the Institute for the Investigation of Barbarian Books, a translation office and the fount of a vigorous industry today. Japan now has over 100,000 professional translators, compared to between 2,000 and 3,000 in the United States.

In 1868 occurred one of the signal events of the 19th century: the Meiji Restoration. It was really a revolution, but the Japanese, with their fondness for softening, cast it as a gentle retreat to the past. It has no parallel in modern history.

After Commodore Perry, and sensing both the menace and boons of modernity, reformers toppled the gasping, distended shōgunate and "restored" the Meiji emperor. The emperor remained a figurehead—like all other emperors for the past 600 years—and real power passed to the reformers themselves.

The shift in power was just the beginning. The Charter Oath of 1868 had five articles. Article 5 stated: "Knowledge shall be sought throughout the world so as to strengthen the foundations of imperial rule." It was as if the preamble to the U.S. Constitution called on American citizens to scout the globe for good ideas. After centuries in the cloister, the Japanese now emerged into the world, blinking their eyes and carefully noting all they had missed.

The Meiji reformers sent observers to the West at once. Japanese studied the French army and German law, universities, and medicine, and they copied the British navy outright. Indeed, in May 1905, when Admiral Heihachiro Togo (1847–1934) ended the Russo-Japanese War by gutting the Russian fleet in the Tsushima Straits, few Westerners realized he had learned his craft at a British naval academy.

The Great Search of Article 5 goes on, and since World War II it has focused largely on the United States. In 1990, for instance, the Japanese published over 3,000 books translated from English, while the United States published just 82 from Japan.[21] The result has been a one-way mirror, whereby the Japanese peer clearly into the United States and we see Japan scarcely at all.

Much of the Great Search has addressed technology. Far more clearly than Americans, the Japanese understand that technological knowledge is wealth. They have therefore assiduously combed Western research, formed joint ventures with Western companies whose technology they wish to learn, and even bought into such firms.

Many Americans interpret this thrust defensively: The Japanese "steal" good U.S. ideas and reap undeserved profits from them. Indeed, the Japanese have not always been moral paragons in their quest for Western technology, as the many cases of "reverse engineering" and other thefts of intellectual property attest. However, in important ways they have proceeded very admirably: spot a promising idea, improve it bit by bit until it is practical, then sell it. Then keep improving it.

The range of ideas they have seized upon and commercialized is remarkable. For instance, take the VCR. A small U.S. firm, Cartridge Television, Inc., marketed the first true VCR in 1972. However, marketing and manufacturing fiascos beset it, and large U.S. firms passed up the chance to bail it out. Sony, Matsushita, and JVC persisted, slashed the device's size and price, and placed it in the home.

Likewise, Americans George Devol and Joseph Engelberger pioneered industrial robots. A self-taught inventor, Devol filed the first patent for a robot in 1954 and tried vainly to sell the idea for two years, until he met Engelberger at a cocktail party. The two formed a company, and in 1962 issued Unimate, a two-ton metal arm like a howitzer which could pick up an object and deposit it on a conveyor belt. When Americans showed little interest, they turned to Japan. Today, robots are everywhere in Japan, assembling machine parts, of course, but also making sushi, fighting fires, playing the piano, and even roving about on the floor of Tokyo Bay.

Americans also invented the fax and the flat-panel computer display, while the Japanese commercialized them. Moreover, the Japanese early on saw the wisdom of American W. Edwards Deming's quality control, which helped cut costs and win hosannas for products like the Toyota, even as buyers worldwide harbored memories of Detroit's "planned obsolescence." Many American businesspeople still have not grasped the cost-effectiveness of quality control.

The Japanese have occasionally inspired a new device itself. Federico Faggin is coinventor of the microprocessor, the master chip at the heart of every desktop computer. It seems a quintessential American triumph. "But in fact, the microprocessor was started because a Japanese company wanted to do it," he points out. "And the first application was a printing calculator for a Japanese company. They are able to see a technological opportunity and invest in it."

Thus, when the Japanese came across fuzzy logic in the course of the Great Search, they were prepared for it. They saw its sense, utility, and potential for profit as most Westerners did not.

7

THE SILK TRACKS

■

A man must make his opportunity, as oft as find it.

—FRANCIS BACON

We fuzzy control theorists have brought control theory from seventh heaven to the ground.

—MICHIO SUGENO

■

The sun is setting in Sendai, a sprawling city of over 800,000 in northern Honshu, surrounded by farms and wooded mountains. As night deepens, the downtown plaza at the bullet train station comes alive. Bars of light march across a façade atop a 10-story building, and on another, jigsaw pieces flash on dit-dit-dit to form a blue neon wavescape. Far below, people stream by after work, bundled up against the chill.

Some seek a relaxing bar or restaurant, but most are returning home. They enter the station, descend gleaming stairs, and buy tickets with brown magnetic backs, which they insert in slots that suck them in and whip them out two feet away. They pass through turnstiles and walk down more flights to the platform, where they may wait a few minutes before boarding a train. When it starts, the standing passengers do something very unusual. They ignore the straps.

The Sendai subway system is the most advanced in the world. On a single north–south route of 13.6 kilometers (8.4 miles) and 16 stations, the train glides along so smoothly that standees sway only slightly. In most cars, if twenty passengers are upright, perhaps four grip a pole or strap. A fish tank can travel the entire route with only a gentle slosh from side to side. One can almost forget one is moving.

Fuzzy logic controls this system and has made it possible. It is perhaps the most conspicuous display of fuzzy logic on earth.

THE ZADEH OF JAPAN

As interest in fuzzy logic cooled in the West, it began warming up in Japan. And if there is a single person who rescued fuzzy control from the limbo of unread journals, it is Michio Sugeno ("suh-GEH-no"), the "Zadeh of Japan."

Sugeno is a brisk, dapper man who wears tinted glasses and, despite streaks of gray, looks much younger than his 52 years. Like many intelligent people, he goes at once to the living core of an idea. "Sugeno is one of the smartest people you can find anywhere," says Zadeh, who adds that he is equally comfortable with playing with theory and soldering circuits. "Not too many people have his range. He is an original thinker and a deep thinker. I'd classify him as a truly outstanding person and that's why he's so influential in Japan."

He was born in 1940 in Yokohama. In June 1945, American planes carpetbombed the city, razing 85 percent of it, and Sugeno still remembers those air raids, or *kūshū* (literally "sky-heat"). One afternoon the five-year-old child watched in terror as bombs struck the homes across the street, turning them to howling flame and later acrid black ruins. He never forgot it, and today, on the walls of his office amid bookshelves and greenery, hangs a print of Picasso's *Guernica*.

Although he enjoyed Western novels as a youth, by high school he had become interested in theoretical physics. He enrolled in the University of Tokyo—the pinnacle of Japan's higher education system, its Harvard, Yale, and Princeton in one. After graduating in 1962, he worked for three years at Mitsubishi Atomic Power, where he was an outspoken Marxist in its labor union. "I wanted to become a member of the Japanese Communist Party, but they rejected me because I was a Trotskyite," he muses. "I was so radical in my student days."

But communism came to disappoint him. It treated the world from a purely rational point of view, he says, much as Descartes dealt with philosophy. It was dry, skeletal, simplistic. Society and life itself were far too rich to succumb to a cry like "Contradiction!" He began shifting over "from rational science to somewhat humanistic science, from Descartes to Pascal," and placed greater value on the subjective.

In 1965 he became a research associate of Professor Toshiro Terano in the control engineering department of the Tokyo Institute of Technology. He first read Zadeh's 1965 paper in the late 1960s and it gripped him. Here it was, he thought, a more subjective rationality, described clearly and sensibly. The paper also convinced him that fuzziness would lie at the heart of artificial intelligence. It hastened his philosophical transition.

When Terano commenced his working group on fuzzy systems in 1972, Sugeno became its secretary. The group met monthly to discuss an array of topics and in its initial year organized the first symposium on fuzzy systems in Japan. That year too, Sugeno asked each member to present his vision of the fuzzy system of the future. He himself described a fuzzy computer, a machine possessed of certain human powers enabled by fuzzy logic.

Yet Sugeno was somewhat dissatisfied with the fuzzy set concept, feeling that it did not lend itself to the applications he was probing. He wanted to solve the problem of the best next move in board games such as chess and go, as well as two favorites from his youth, poker and mah-jongg. Some researchers were then using probability theory, but Sugeno sensed it was a cul-de-sac. Hence, he invented the theory of fuzzy measures, which he articulated in his 1974 Ph.D. dissertation.[1]

Fuzzy measures assess how much the evidence proves a fact. Suppose there are ten balls, each with a number from 1 to 10 beneath it, and a problem-solver must guess the numbers using hints, without looking at the balls. Fuzzy measures describe how certain one can be about each guess. For instance, if the clues suggest that a ball is a 7, the degree to which the evidence proves it so—say, 0.6—is a fuzzy measure. Similarly, if a patient were suffering an unknown lung ailment, a physician might deem it either bronchitis, emphysema, pneumonia, or a cold and assign the following fuzzy measures: bronchitis, 0.75; emphysema, 0.45; pneumonia, 0.30; and cold, 0.10. In probability, such figures would have to add up to 1, no matter how much or how little proof there was. In fuzzy measures, they don't. Rather, each measure starts at zero and works up. Likelihood is quite different from degree of evidence.

In 1975, Sugeno met Ebrahim Mamdani at a symposium in Boston, and later Mamdani invited him to Queen Mary College. Sugeno stayed there for eight months starting in December 1975 and became acquainted with fuzzy control. But at the time Mamdani's work failed to interest him.

"I'll tell you something," he says. "I was a control engineer, so I didn't think it was possible to apply fuzzy set concepts to control engineering. Control engineering is so complicated." Mamdani, on the other hand, strode right in because he "didn't know anything about control. That's my conclusion. He was very brave, you know."

After Sugeno left England, he spent another eight months in France and in 1977 returned to Japan. He was becoming more and more absorbed in fuzzy theory, but found few people paying heed to it. He finally realized he could best spur interest with practical success stories. Hence, in 1978 he ceased researching fuzzy measures and shifted to control.

He began by creating Japan's first fuzzy application: control of a Fuji Electric water purification plant.

Water purification is a complex process. One first takes water from a lake or river and places chemicals like aluminum sulfate in it to clarify it. These agents are coagulants. They bring the impurities together into flocs, which are heavier and sink faster. This step is vital, because if dirty water moves on to the next step, filtration, it clogs the filter and flow simply stops.

However, the level of impurities in water is constantly changing, so the rate of which the plant releases coagulants into the water must vary as well. For instance, rain erodes soil and temporarily dirties the rivers. But that is not all. The rate must also change to reflect water temperature, alkalinity, floc size, and other factors. The whole relationship is intricate and defies simple modeling. Moreover, sedimentation takes about four hours from start to finish, so feedback arrives too late to help. It therefore requires foresight, knowledge. Hence, control of the process had always required an experienced individual. Fuji Electric wanted to automate it.

"We observed the human operator's action," Sugeno says. "We took input data, which we watched, and we also corrected his output: the amount of chemical to be put in the water." He then devised about ten fuzzy IF–THEN rules to simulate the operator's actions and placed them in a system like Mamdani's. The results matched the operator's and clearly exceeded a crisp approach. It had taken three years, from 1980 to 1983. "That was my first experience," Sugeno says. "After that, we could shorten the time."

In 1983 he began pioneer work on a fuzzy robot, a self-parking car which he controlled by calling out commands. It was a miniature vehicle, and came equipped with a Ricoh speech recognition device.

His first problem lay in the instructions themselves. What did it mean to say, "Turn right"?

In fact, he realized, "turn right" means to turn right if and when it's possible. Tell a taxi driver to turn right and he doesn't instantly swing the steering wheel. He waits until the first street. Thus, this simple utterance carries hidden assumptions. It really means: "Go straight until you reach a corner, then turn right." We automatically read this extra information into it because of our shared experience and knowledge of context, that is, our common sense. We know the taxi driver will wait till the corner to turn, so there is no need to tell him to. Machines lack such background and cannot create it spontaneously. Sugeno had to endow them with it. He had to give them common sense.

So he applied a fuzzy algorithm which included this vital mote of knowledge and, in fact, expanded the meaning of "turn right" still further. A taxi driver does not suddenly jerk the steering wheel 90 degrees at the corner. Instead, as the car approaches, he brakes gradually and at the point of the corner steers slowly to the right. Then, before ending the turn, he turns slowly left and speeds up. Sugeno taught his little car to do likewise. He worked on this problem for another three years and eventually could park the vehicle by simply asking it to park.[2] It was an impressive feat and, for Sugeno, it was just the beginning.

THE SOUL OF A REALLY NEW MACHINE

As Sugeno pioneered fuzziness, awareness of it in Japan slowly grew. Tellingly, one of the few individuals making progress in the United States was a Japanese-born naturalized American named Masaki Togai.

Togai and Hiroyuke Watanabe invented the first fuzzy chip, a device that promised great increases in speed and power. For instance, while Sugeno's water purification plant needed about 10 rules, Lotfi Zadeh says a fuzzy chip can hold 256 or even 1,000. "And the rules are easy to enter," he says. Yet when Togai and Watanabe tried to persuade their American employers to market the chip, they met stifled yawns and polite dismissals.

Togai is a genial, compact man with a bright smile and bangs like the early Beatles. His father was a flight engineer, and when the F86 became part of the Japanese air force, was one of the first Japanese to visit the United States to learn how to fly and maintain it. He brought back many exotic gifts to the family, such as oatmeal, a novelty to the young Togai. When he was six or seven, his father returned with a mechanical pencil sharpener. "That small, well-made pencil sharpener was an eye-opener for me," Togai says. "It was my first opportunity to admire a U.S. product."

He entered Japan's National Defense Academy and after graduation became an aircraft maintenance engineer. He stayed in the air force for six years. The first three were arduous. He had to awake at six in the morning, prepare an aircraft to fly, then labor on other planes till it returned. Sometimes there were three flights in a day and he worked till 10 P.M. In his third year, he followed this schedule while also studying for his graduate school entrance exam. "I was working without sleeping," he says.

But the exam went well. Each year the Japanese air force sent two individuals to graduate school in the United States, and that year it selected Togai. On the recommendation of one of his advisors, he decided to attend Duke University, in North Carolina. "I didn't even know where North Carolina was," he laughs.

He arrived at Duke in 1975 speaking little English, one of only two or three Japanese students on campus. "My first impression of North Carolina was that it was the middle of nowhere," he says. "Pine trees, pine trees, nothing but pine trees." The first few weeks he didn't even know what an "assignment" was, and sensed he might have missed something when he saw people submitting work.

However, he came to enjoy it. The campus was beautiful and the students were friendly. He had trouble communicating with them at first because of the Southern accent. "I was not the only one," he says. "The people from Boston also had a problem."

At Duke, his advisor was Paul Wang, and he studied under Peter Marinos. Both were from Bell Labs and both were interested in fuzzy logic. In one class Marinos discussed it, but Togai saw no future in fuzzy logic and disregarded it.

Armed with his master's, he returned to Japan and the air force in 1977, working in the Research and Development Institute of the Defense Agency. There he designed an early warning receiver, an "electronics warfare gadget" for the F15. It sought to analyze incoming radar, determine its type, and warn the pilot so he could evade the enemy craft. It was a problem in pattern recognition, and Togai was using traditional logic. It didn't work.

"If you're classifying these things using a yes and no approach, you wind up with nothing, so it says, 'Unknown,'" he says. Why? In most cases, the receiver has only several pulses to analyze, and they just don't yield enough data. "So if you use the conventional approach, it's almost impossible. What it can tell the pilot is: 'Something is aiming at you.' That's it." Much later, after founding his own firm, he approached his former vendor, Tokyo Keiki, and suggested using fuzzy logic. They agreed, and Togai gave them more accurate results.

Meanwhile Paul Wang had written him a long letter urging him to return for his Ph.D. Wang promised a raft of intriguing projects and the stipend exceeded Togai's salary in Japan, so he re-enrolled at Duke in 1979.

By now he was more interested in fuzziness. He worked at Duke's huge medical center, and one of his first projects involved mice and drugs. He had to guess how much drug experimenters had injected into a mouse by observing its deportment: running around, sniffing heavily, dozing. The task was highly subjective. "It depends on how you feel, how you see it," he says. So they used fuzzy sets to classify responses, and "compared to the conventional statistical method, it gave better results."

He also worked for Lord Company on a tactile sensor for robots, a device that classified objects by touch. For instance, if a person sticks her hand in her pocket, she can more or less tell whether she is grasping a dime or a quarter. Togai and his colleagues wanted to give robots that power, and fuzzy logic came in handy for two reasons. First, Togai says, "The boundary is kind of fuzzy. That's the reason we started using fuzzy. It says, 'This is more likely a quarter, but it could be a dime.' Then you can change. You can measure it again to confirm it." Moreover, the tactile sensor often yielded precious little information and fuzzy logic, like a kind of Sherlock Holmes, could take it a long way. Lord Company eventually issued the sensor as a product.

He did his thesis on fuzzy inferencing. In it, he proposed the architecture of a fuzzy inference engine, a fuzzy hardware device to process inferences at high speed. It was the theoretical foundation of his fuzzy chip.

In 1982, John Jarvis, head of the robotics research department at Bell Labs, came to Duke recruiting. Togai was teaching Japanese, and one of his students, the wife of a physics professor, introduced him to Jarvis. Jarvis was intrigued by Togai's work on tactile sensors and hired him. In 1983, he went to work at Bell Labs in Holmdel, New Jersey.

Bell Labs was long America's premier research institute. In 1947, William Shockley, John Bardeen, and Walter Brattain invented the transistor there, the key ingredient in radios, TVs, and computers. When Togai arrived, it abounded in Nobel Prize winners. For instance, vice president of research Arno Penzias, who won for detecting the red-limit radiation left over from the Big Bang, occasionally dropped by the lab for bagels in the morning. Togai joined Bell Labs around the period of divestiture, and the number of researchers had dropped from around 50,000 to about 20,000. Even so, Togai says, "Holmdel was a huge building, with about 4,000 people there. It was like a city."

Togai worked in the computer system research division, trying to fashion an intelligent robot that could play ping-pong. The machine had to learn and react quickly. It needed good mechanical systems and vision, and it had to be portable. The first robot ping-pong tournament was later proposed in 1987, and Togai says there were no entries. "The best you can do is serve the ball," he notes. "You can't hit it back, which is the hard thing."

To achieve this feat, he and Hiroyuke Watanabe began designing a fuzzy chip. Speed was a critical factor. A conventional binary chip can perform fuzzy logic, and it will work fine in many cases. A fuzzy chip, however, greatly accelerates fuzzy operations. It can perform in a single pass what a digital chip requires several passes to do. Moreover, as Togai points out, it cuts the steps in writing the program.

A digital chip is a series of switches based on binary logic. The central idea is simple. For instance, two of the more important operations in binary logic are the AND (e.g., "The sky is blue AND the grass is blue") and the OR ("The sky is blue OR the grass is blue"). An AND statement is true if both its parts are true, and resembles two bridges in a row. If one bridge collapses, a car can't get through. In logical terms, the AND is false. An OR statement is true if either part is true, and resembles two bridges across the same river. If one collapses, a car can cross on the other one. Both must fall to stop the car, or to yield a false. Replace the road in the analogy with a path for electric current and the bridges with switches, and the result is an electronic logic operation. Link millions of them together, etch them on a chip, and a computer emerges.

But at bottom digital computers rest on true/false logic. The current either crosses or it doesn't. Fuzzy chips work differently. In his seminal 1965 paper, Lotfi Zadeh described their basic architecture. He noted that while switches can represent binary logic, "sieves" could model fuzzy logic. Instead of admitting or stopping current, they could let different levels of current pass. By changing the "mesh size," one could perform fuzzified logic operations. For instance, a fuzzy AND would choose the smaller of two current levels, exactly as a committee member in a fuzzy system recommends the lesser of two memberships in, say, negative small pressure error and zero pressure change.

Togai and Watanabe presented their paper in December 1985 at a Miami Beach conference on AI. After this event, many Japanese firms such as Fuji Electric notified Togai that they wished to place the fuzzy chip in their systems. So Togai and Watanabe approached Bell and suggested that they commercialize the chip. Bell rejected this idea. The pair were in research, not development. So they approached development directly, but no one showed any interest. "It was quite new to them and at the time conventional AI was booming," Togai says. "The AI community is still very hostile to fuzzy."

"It was looked on as oddball," says Bell Lab researcher Alan Huang. "It was soft science, not quote traditional. You'll see that fuzzy logic also didn't take off in certain academic universities like Stanford or MIT. It's the same symptom. It's not just Bell Laboratories. If you're not part of the party line . . ." He shrugs. "Some things just get a bad rap and you're a heretic."

In 1986 Togai and Watanabe left Bell Labs and joined the Science Center of Rockwell International, in Thousand Oaks, California. There they designed a second chip and also did work for NASA, which had by then become involved in fuzzy logic. This new chip sparked further interest from Japan, and representatives of several Japanese firms flew in to meet with Togai and Watanabe. The pair again sought to market the chip, but Rockwell's vice president in charge of engineering declined. They responded that, if Rockwell didn't, the Japanese would build it themselves. Fine, said the vice president. Rockwell would buy from them. "There was no hope," Togai sighs. Rockwell was an outsourcer, he says, and bought from other firms wherever possible. "That has been their mentality." Togai grew restless and started casting about for something to do.

THE IRON MAN

Togai was right. A scientist in Japan was already working on fuzzy chips. He was Takeshi Yamakawa, another remarkable character in the fuzzy world. "He is a man of vision," says Zadeh. "I've met lots of people who are extremely smart and capable, but I don't bump into people that frequently who have vision." Says Bart Kosko, "Yamakawa is like a samurai. He's a charismatic figure."

He was born in 1946 in Shenyang (Mukden), Manchuria, where his father had been a bank director in World War II. When he was three months old, his parents returned to Japan, and his mother died six months later, "so I don't know my mother's face." He grew up in his grandmother's house in Kumamoto, on the warm southern island of

Kyushu, and wanted to be a physician. However, he failed the entrance exam twice, the second time because he misread a question. "That was the change of my life," he says.

But Yamakawa does not give up easily. A karate expert, he has carefully read Musashi Miyamoto (1584–1645), author of *The Book of Five Rings*, the famous tract of strategy and advantage. "Musashi was a very great soldier," he says. This fearless samurai studied the craft of fighting and mastered it, and went on to study the way of living, to become a warrior-philosopher. Near the end of his life, he became a well-known Japanese brush artist. "So if a person can achieve one thing," Yamakawa says, "that way will lead to anything."

He shifted over to electronics. He had enjoyed electronics since elementary school, built radios and stereos in junior high, and at 17 won an award for his design of a stereo amplifier. He graduated from the Kyushu Institute of Technology and took his Ph.D. at Tohoku University in Sendai. Medicine, especially neurophysiology, still fascinated him, and in 1969 he wrote about electronically modeling how the brain fuses signals from the two ears. He also began studying frog neurons and soon immersed himself in electrochemistry, since all brain cells carry information electrochemically. In particular he probed inert metals such as gold and platinum, and wrote his Ph.D. thesis on their deposition. This knowledge proved useful later when he began making chips.

After graduation, he remained at Tohoku University three more years as a research assistant, then moved down to Kumamoto University on Kyushu, where he worked on motors, generators, transformers, and other electrical devices. He would later apply this background to chips as well.

He first heard of fuzzy logic in 1977 and soon found himself devising independent fuzzy logic circuits, the building blocks of chips. Fuzziness was still largely unknown in Japan. "When I was a young associate professor, many people asked me, 'How is fizzy?' Not 'fuzzy,' but 'How is fizzy?' " In Japanese, "fizzy" is Fiji. In 1979 he wrote a paper about his work. The referees rejected it. They said no one could ever combine Yamakawa's tiny parts into working silicon chips. He had to prove them wrong. "I wanted to fabricate the integrated circuits by myself, but I didn't know to do it," he says.

So he flung himself into the problem. Yamakawa is a man of fabled stamina who has trained himself to sleep three or four hours per day. "It's not so difficult," he says. "At 8:30 A.M. I will be in my office, and I will be back home at 3 A.M. or so. That is daily life." When working effectively, he "forgets" to sleep, and in one burst of energy in 1990, he stayed awake for 60 hours, then delivered a lecture to his class and attended an international conference.

He obtained a room, divided it in half, and set up a fabrication line. After six months, he finally succeeded in assembling larger fuzzy integrated circuits. But his makeshift factory remained too primitive to create a total system. So in 1986 he asked his dean where he could get help. The dean suggested Omron Tateishi Electronics, in Osaka.

Omron is the linchpin of much of Japanese electronics. It developed or helped develop some of the first pocket calculators, as well as ticket vending machines, traffic monitors, automatic tellers, bill changers, point-of-sale terminals, healthcare equipment, artificial limbs, and, recently, a UNIX workstation. "Tateishi is today the hidden manufacturer which enables other Japanese manufacturing giants to function," says one writer. "It is Tateishi component parts which operate the assembly lines of other companies."[3]

Founded in 1933 by an inventive engineer named Kazuma Tateishi, who spent its first few years peddling products on a bicycle, Omron remained fairly small for two decades. Then in 1952 he entered cybernetics and began making control relays for automation. It was his master stroke, the start of serious corporate growth, and he never forgot the importance of listening carefully to new ideas.

Omron played a pivotal role in fuzzy logic in Japan. It had begun its fuzzy R&D in 1984 with a fuzzy expert system for diagnosing illnesses, which it completed in 1986. (By September 1990, according to figures it presents, Omron held 107 of the 389 fuzzy patents in Japan.) Hence, the firm Yamakawa approached knew more about fuzzy logic than perhaps any other in the world.

"I asked them to construct or fabricate the integrated circuit designed by us," Yamakawa says. The company agreed, and Omron began selling the first dedicated fuzzy controller in 1988. The same year Omron established its F-project team, and engineers and marketing people confected further fuzzy chips designed by Yamakawa.

THE SILKEN RIDE

In the rolling hills of Kawasaki, a city of over 1 million outside Tokyo, one approaches a tall mesh gate, speaks for a moment with a smiling guard, then passes through. For a moment, one gazes out over the vast Kantō metropolis, the largest urban area on earth, then descends a curving asphalt road into a green, wooded dell. It is the site of the Hitachi Systems Development Laboratory, one of the seedbeds of the future.

The lab is part of a massive R&D organization. Hitachi is the second largest electronics firm in the world and maintains 22 laboratories in Japan and 4 overseas. These labs probe an astonishing array of topics, including superconductivity, language translation machines, neural computers, and high-definition TV. How big is Hitachi's commitment to research? In the year ending March 1991, the company pumped $3.48 billion into R&D—more than the GNP of such nations as Jamaica, Liberia, and Madagascar. In comparison, in 1990 the NSF spent $2.08 billion, about three-fifths the Hitachi sum.

Here in 1979, ensconced in the glen, researchers Shoji Miyamoto and Seiji Yasunobu began studying how fuzzy logic might help control the Sendai subway. It was a large, complex task. "We were not so confident at first," says Miyamoto, a quiet man of 46 with a warm smile, "but we thought fuzzy logic could be very useful."

However, they quickly perceived a problem in mastering the rapid pace of events. A subway is a real-time system, so any controller must regulate change as it occurs. Yet fuzzy logic was lagging behind. It seemed too slow to control the train.

In the early 1980s, Yasunobu saw the key. The fuzzy system was reacting to events rather than anticipating them. For instance, the train would begin a descent and accelerate, and the system would then brake. But human operators knew the route and slowed before reaching the downhill. Yasunobu decided they should equip the train with the same kind of foreknowledge, and devised predictive fuzzy control. They taught it the route. Thus edified, it would slow while nearing a downslope and speed up before reaching a hill. This idea proved the turning point, and by 1983 the two men were convinced the fuzzy subway would work.[4]

However, they faced not just technical but regulatory challenges. "The Japanese government is very conservative," says Zadeh. "They had to persuade many people that this was the right thing to do. So it was a gamble." Before they could start operating the train, Miyamoto and Yasunobu had to conduct hundreds of thousands of computer simulations and some 2,000 actual runs on the track.

The Sendai system officially debuted on July 15, 1987. Two days later, Zadeh arrived and the Hitachi engineers showed him around. He viewed the control center, the repair facilities, the documentation of the countless trials. He also rode the train. "It was gratifying that the basic ideas underlying fuzzy logic had finally found their application," he says. "The whole system was highly impressive."

This subway has the smoothest ride on earth, but other virtues as well. It can stop almost on a dime, within seven centimeters of target, three times better than a human operator. It also saves energy, cutting costs a whopping 10 percent. Moreover, says Zadeh, "They've never had a mishap." Such virtues convinced the Tokyo municipal government to hire Hitachi to install fuzzy logic on part of its labyrinthine subway system.

THE BIG SHIFT

The floodgates opened up for fuzzy logic on July 18, 1987, at the second annual International Fuzzy Systems Association (IFSA) conference in Tokyo. Prior to this event, fuzzy logic was not well known in Japan. After it, a wave of pro-fuzzy sentiment swept through the engineering, government, and business communities. It created a consensus, a general (though not unanimous) feeling that fuzziness repaid scrutiny.

The conference was splendidly timed. It began three days after the Sendai subway opened, and attendees were abuzz with its dreamy ride. But Professor Kaoru Hirota of Hosei University also displayed a fuzzy robot arm that played two-dimensional ping-pong in real time.[5] And Takeshi Yamakawa unveiled his fuzzy controller at this conference, and it was a sensation.[6]

Yamakawa demonstrated a fuzzy system that balanced an inverted pendulum, that is, a pole that can fall only left or right. "It is very, very difficult to realize that using the traditional computer. But the fuzzy controller achieved it very easily," he says. Moreover, he was able to place a flower atop the pole and, to the astonishment of engineers, balance that as well.

As he was exhibiting the inverted pendulum, one spectator asked him to remove a board from the fuzzy computer. "I thought that was incredible, crazy," Yamakawa says. If he detached a board, the pole would drop at once. But the man persisted, saying he wished to see exactly how it would fall. So Yamakawa disconnected a board and, to the surprise of everyone, the pole remained upright. The controller continued to regulate it. This serendipitous demo showed the graceful degradation of fuzzy systems, their ability to make decisions even with partial information.

The inverted pendulum vividly displayed the power of fuzzy logic. It had immediate impact on companies, and *Asahi Shimbun*, one of the three giant Japanese dailies, ran a front-page article on it. "That was the trigger to spread the fuzzy logic systems to industry," Yamakawa says. After that, he gave more than 150 lectures in one year.

The IFSA conference had one more hallmark. The first fuzzy consumer product debuted there, a small device that controlled a shower head. "That came as a great surprise," says Zadeh. "Nobody at that time expected a consumer product."

8

IN ELECTRIC TOWN

■

Some day the enormous amount
of effort the Japanese have put
into fuzzy logic will earn Lotfi a
medal that will be awarded by
the President of the United States.
God knows we have to do something to slow those guys down.

—WILLIAM KAHAN

It is not the going out of port, but
the coming in, that determines
the success of a voyage.

—HENRY WARD BEECHER

■

Akihabara, Tokyo's "Electric Town," is like no other neighborhood on earth. Beneath ten-story structures hung with brightly hued banners lies an electronics bazaar. Narrow arcades burrow through buildings, lined with shop after shop selling only electronics. No restaurants, no travel agencies, no grocery stores, no pachinko parlors. It is all technology.

And the newest consumer electronics in the world often surface here first.

The browser in these high-tech souks sees karaoke machines, digital saxophones, roadmap machines, handheld transceivers, digital drums, electric shavers that look like mushrooms or chic cigarette packs, TVs with sharp two-inch screens. And every once in a while, the subway roars past, four stories above.

Akihabara is the showcase for the newest fuzzy hardware in the world, the fuzzy washers, camcorders, and microwaves. By late 1990, at least 20 kinds of fuzzy consumer products had debuted. Some stores displayed them at the high-end of the line, with the legend "FUZZY" splashed across them, and other shops set aside a special corner for fuzzy products.

For retailers, fuzzy means money. At Dai-Ichi Katei Denki Company, an Akihabara appliance shop, a spokesman said fuzzy products had "captured our customers' hearts" and kept him "terribly busy every day." Fuzzy washing machines were selling best, he noted, followed by rice cookers and vacuum cleaners. An employee at Shintaku Electric Appliances added, "Customers seem to believe anything fuzzy-controlled is a good product, so we'll continue to promote sales of this kind of goods."[1]

William Kahan may claim fuzzy logic is a boondoggle for the Japanese, but he has never been to Electric Town.

BEYOND DESCARTES

After the second IFSA conference, the first company off the blocks with fuzzy consumer products was Matsushita Electric Industrial

Company, known in the United States as Panasonic, Quasar, and Technics. This firm alone now offers fuzzy washing machines, vacuum cleaners, rice cookers, kerosene heaters, refrigerators, air conditioners, microwave ovens, hot- and cold-water mixing units, video camcorders, and many other products. It has also installed a fuzzy traffic control system in Nagoya, a city of over 2 million.

The company was founded in 1917 by Konosuke Matsushita (1894–1989), whose father on his deathbed told him that, as the last male in the family, he had a duty to make Matsushita a famous name in business. He made it legendary. It is the world's biggest maker of consumer electronics, and in 1990 *Fortune* ranked it the 17th largest industrial firm on earth. It specializes in bringing inexpensive goods to market quickly. It is not known as an innovator, like Sony, but rather as a canny maker and merchandiser of consumer products. "Nobody outmanufactures Matsushita," says *Business Week*.[2]

Much of the key work on fuzziness at Matsushita Electric took place at its Central Research Laboratory in Osaka. The visitor to the lab arrives in an open area between three white buildings, where there stands a striking tribute to Western science: a frock-coated Thomas Edison atop a pillar, circled below by busts of sages of electricity like Ohm and Marconi.

Here, in 1987, Yoshihiro Fujiwara was general manager of research planning. A tall, pleasant man with the look of a scholar, Fujiwara was casting about for a new project in control. He had heard of fuzzy logic through technical journals and academic societies, and sensed its promise. In October, he took his boss, lab director Tsuneharu Nitta, to visit Michio Sugeno in the latter's 11th-floor office at the Tokyo Institute of Technology in Kawasaki.

Sugeno himself had an interest in this meeting. He wanted to start a national project with the Science and Technology Agency (STA) and sought Matsushita's participation. He also wished to place fuzzy control in home appliances, Matsushita's great domain.

Modern science, he told his visitors, sprang from Aristotle and Descartes, who exalted the precise and shunned the fuzzy. But thought is not naturally crisp. "He said we should jump over the Cartesian way of thinking," Fujiwara recalls. Rationality has a human side, Sugeno stressed, and we have long neglected it. He mentioned no particular applications, but he did say, Fujiwara remembers, "You, Matsushita, should implement fuzzy technology in consumer products first in Japan and in the world." With its size and astuteness, Matsushita could forcefully demonstrate the virtues of fuzziness.

Nitta and Fujiwara departed politely but noncommittally, and Sugeno could not read their response. In fact, it was highly positive. "The way he talks is soft-spoken," says Fujiwara, "but he is strong in mind, in heart. And he speaks very passionately. I was impressed by him very much."

As a result of this conversation, Nitta decided to focus on fuzziness. He transferred Fujiwara from his previous post to general manager of the systems and control group at the lab. There, he became head of line research, and Noboru Wakami, research manager of control technologies, headed up the development effort.

They began investigating fuzziness and at once met resistance from division engineers. These qualms were a major obstacle, since the Central Research Lab did not push technology on other labs, but tried to create demand for it. "We didn't say, 'You must use this,' " Fujiwara says. "We just said, 'This is very useful. If you agree, please introduce it.' "

To win them over, the lab built a fuzzy robot vacuum cleaner for Matsushita's 70th anniversary technology fair in 1988. This beetle-shaped device, slightly larger than a wastebasket, scooted smoothly across the floor and cleaned the entire room without supervision. "The engineers of division laboratories were very influenced by the results of the vacuum cleaner robot," says Fujiwara. It had particular impact in the Matsushita home appliances laboratory, whose director, Junotaro Mikami, began urging others to develop fuzzy appliances. Soon engineers were working on fuzzy products.

At the same time, Wakami was developing a fuzzy expert system shell, a core program which anyone could use to create a fuzzy appliance. "We transferred the shell to the divisions as a development tool," Fujiwara says. "That was the beginning of consumer appliances."

Matsushita introduced its Aisaigo (beloved wife's machine) Day fuzzy washer in February 1990. It was the first major fuzzy consumer product, and it brought one-button intelligence to washing machines, automatically setting the proper cycle according to the kind and amount of dirt and the size of the load.

Like the other fuzzy appliances in Japan, it basically follows the model of Mamdani's steam plant and the Smidth kiln. It uses fuzzy

IF–THEN rules to merge information from three inputs—amount of dirt, type of dirt, and load size—into one output, the correct cycle.

Sensors supply the Aisaigo Day with the inputs. Its optical sensor sends a beam of light through water in a pipe and measures how much reaches the other side. The dirtier the water, the less light crosses. As the wash cycle commences and dirt dissolves in the water, it grows darker and darker. Less and less light strikes the sensor, until at some point the water becomes as opaque as it will get. That point indicates how dirty the clothes are.

The optical sensor can also tell whether the dirt is muddy or oily. Muddy dirt dissolves faster. So if the light readings reach minimum quickly, the dirt is muddy. If the downswing is slower, it is oily. And if the curve slopes somewhere in between, the dirt is mixed.

The machine also has a load sensor, which registers the volume of the clothes.

While the steam plant required ongoing feedback, the washer's inputs, once known, are fixed and final. The amount of clothes, for instance, will not change over time. The controller merges these fuzzy sets and defuzzifies, not into numbers, but rather into one of six hundred processes. Each involves a different mix of water volume, strength of flow, and washing time. Matsushita claims the fuzzy washer eliminates such problems as damage to laundry from excessive washing and poor performance from insufficient washing. It also saves time and energy.

Fujiwara says Matsushita has been manufacturing 35,000 fuzzy washers per month from the start, twice the number of the previous model. "We are limited by factory capacity," he says. "We are producing as many as we can."

Sanyo and Hitachi also have one-touch fuzzy washers. However, each addresses different factors. The Sanyo device adjusts for how well the detergent has dissolved in the water. The Hitachi judges the weight and type of fabric, and uses fuzzy logic to set the proper washing cycle, with emphasis on not damaging the fabric over time.

IN THE HOME

The washer has generally received the spotlight among fuzzy appliances, but others abound and more are springing up all the time. For instance, Matsushita and Hitachi sell fuzzy vacuum cleaners, which sense the amount of dust on a carpet or the condition of the floor and set the proper suction level.

Dust does not collect evenly in a room, but falls more heavily where people tend to gather, such as near the telephone, tables, and chairs. Moreover, a pile carpet will accumulate more dust than a hardwood floor. Matsushita's fuzzy vacuum cleaner automatically adjusts for these areas and thus speeds cleaning, operates more effectively, and saves power.

It is a simpler version of the washing machine. An infrared sensor detects how much dust is passing into the machine. The sensor relays this information to a microprocessor, which selects the proper power for the vacuum cleaner. The more dust, the higher power. Because the sensor also tracks the changing level of dust, the vacuum cleaner can tell the user how much remains on the floor. Three red lights indicate plenty; two, some; and one, little. When the floor is clean, a green light comes on. This feedback, the company says, is especially handy for cleaning under tables and other places out of view.

Matsushita and Sanyo make fuzzy rice cookers, which allow one water level to yield four different kinds of cooked rice: hard, medium, soft, and sushi. "Without fuzzy logic, we have to change the water level," says Fujiwara. Soft rice, for instance, might require more water. Fuzzy logic enables the device to choose among different methods of steaming, so the user can fill the pot to a single marked line in every case. It is a simple convenience, but, he says, it "is not possible without fuzzy logic."

Masaki Togai's firm did the software for a fuzzy air conditioner from Mitsubishi Heavy Industries. The company says it saves 24 percent in energy costs over its traditional PID controller when cooling and 17 percent when heating. "As a matter of fact, my parents are using fuzzy air-conditioning," Togai says. "They say their electrical bill is far lower than last year even though it's been a hot summer." The device also recognizes when people have entered the room and cools more vigorously.

Sanyo, Sharp, and Toshiba sell fuzzy microwave ovens, which select the proper time and intensity for cooking foods. According to Togai, to avoid overcooking, the fuzzy microwave has sensors for temperature and humidity.

ON VIDEOTAPE

The human eye is continually jumping around. It moves its focus automatically about five times every second in a random perusal of the visual field. These abrupt shifts are called *saccades*, and most people are surprised to learn of them. We simply do not perceive them. Through processes no one fully understands, the brain edits this jiggle out, so the world appears constant and stationary.

Matsushita faced a similar editing task when it developed a small one-hand camcorder. A leather strap binds the camcorder to the hand, and the photographer then aims the device at the target. But a hand is not a tripod. It proved impossible to hold the camcorder without shaking it slightly and imparting an irksome quiver to the tape.

Smoothing out this jitter was tricky. "We had to distinguish the motion of vibrating hands from moving objects in the picture itself," says Fujiwara. "When my hand is shaking all of the points in the picture are moving in the same direction at the same speed." But if an object like a car crosses the field, only some points will change. Therein lay the key.

To identify such localized motion, Matsushita developed what it calls a digital image stabilizer, based on fuzzy logic. The stabilizer rapidly compares each current frame with the previous images in memory. If the whole appears to have shifted, the fuzzy controller adjusts the frame to match. Otherwise, it leaves it alone. Thus, if a person walks into view, the image changes on just one side, so the camcorder doesn't try to compensate. Even if an earthquake struck and the whole scene started to tremble, the camcorder would not attempt to correct it. "Individual things like flowers in a vase would shake in the picture, which is different from hand jitter where everything is moving at the same time," Fujiwara says. Technically machines cannot identify true jitter in every case, but the exceptions are rare enough to make this approach very practical.

Matsushita introduced the fuzzy camcorder in June 1990. In December 1990, *Fortune* named it a Product of the Year. In the same month, Gene Siskel compared the Panasonic and the Sony handheld camcorders on the TV show "Siskel and Ebert." He preferred the Sony, given its array of fine qualities, but said the antijitter feature of the Panasonic impressed him most. It alone, he said, almost raised the Panasonic to the level of the Sony.

This product too has caught fire in the marketplace. Initial production ran 35,000 per month, and by February 1991, eight months after introduction, it had risen to 90,000 per month.

UNDER THE HOOD

Fuzzy logic has also found its way into cars, where it is proliferating. An automobile is a collection of many systems—brake, air-conditioning, suspension, transmission, steering, and more—and fuzzy logic can enhance almost all of them.

For instance, Nissan has patented a fuzzy automatic transmission which, according to Togai, cuts fuel costs by 12 to 17 percent. A normal transmission is crisp, shifting whenever the driver passes a certain speed. It therefore changes quite often, and each shift eats up gas. However, human drivers not only shift less frequently, but consider nonspeed factors as well. If accelerating up a hill, for instance, they may delay the shift. Nissan's device changes gears about as often as a driver with a manual transmission.

When it comes to saving gas, 12 to 17 percent is a huge amount, both per individual car and for society overall, and it is especially significant in nations like Japan that must import their oil. "That's why Honda is very much excited about it," Togai says. His firm has worked with Honda developing the rules and software for this device. Subaru has likewise placed a fuzzy transmission in its Justy.

Nissan has patented a fuzzy antilock braking system. The challenge here is to apply the greatest amount of pressure to the brakes without causing them to lock. The Nissan system assesses 18 different factors. For instance, if the car slows down very rapidly, the system assumes brake-lock and automatically eases up on pressure.

In April 1992, Mitsubishi announced a fuzzy omnibus system that controls a car's automatic transmission, suspension, traction, four-wheel steering, four-wheel drive, and air conditioner.[3] The fuzzy transmission, for instance, prudently downshifts on downgrades or curves, and also keeps the car from upshifting inappropriately on bends or when the driver releases the accelerator. The fuzzy suspension contains sensors in the front of the car that register vibration and height changes in the road, and adjusts the suspension for a smoother ride. Fuzzy traction prevents excess speed on corners and improves the grip on slick roads by deciding whether they are level or sloped. Fuzzy steering adjusts the response angle of the rear wheels according to road conditions and the car's speed, and fuzzy air conditioning monitors sunlight, temperature, and humidity to enhance the environment within the vehicle. At the time of the announcement, Mitsubishi predicted the all-purpose system would soon appear in a compact car.

THROUGH THE LENS

Canon deserves credit for the biggest oxymoron in the field so far: fuzzy focus. This company began in 1933 when physician Takeshi Mitarai learned that a Leica camera cost as much as a house in Japan and decided he could undersell it. He founded his firm in the now-chic Roppongi district of Tokyo and in 1934 issued his first camera, initially called Kwannon after the Japanese goddess of mercy, but renamed Canon even before its release. The company later prospered in X-ray cameras and automatic eyes, and in the 1960s struck another rich vein in the copier business.

Today, Yasuteru Ichida is general manager of the circuit technology division of Canon's Production Engineering Research Laboratory in Tokyo. He is an animated, pleasant man who speaks cheerfully about the search to find the best techniques for electronics products. He first heard of fuzzy logic in 1987 at the second IFSA conference, along with his group of five engineers. A highlight was Takeshi Yamakawa's self-balancing pole. "That had a very big impact," says senior research engineer Yuzo Kato, a member of the group. Several hundred people witnessed the demonstration, and Kato says control engineers "were very surprised at the simple way fuzzy logic achieved the results." When Yamakawa cut part of a flower and placed it atop the stick, says Kato, "That shocked the control engineers." Ichida was also impressed. "It realized pendulum control in a very simple and inexpensive way," he says. "If you want to do that using conventional technology, it takes a huge amount of money." And Takeshi Sawada, general manager of the Applied Information Technology Division, took special note of the fail-safe feature which Yamakawa had involuntarily demonstrated.

Not everyone succumbed. Development engineer Kitahiro Kaneda is a young, casually dressed engineer who earned his master's in mechanical engineering in 1986 and then joined Canon. He didn't see the inverted pendulum, but the reports failed to impress him. However, soon he heard of fuzzy engineers balancing several rods at once. That was remarkable, he felt. It brought home the potential of fuzzy technology.

Kaneda first worked in mechanical parts development for video products, but he was interested in software engineering and after a year, during a lab restructuring, asked for a new assignment. The firm gave him autofocus technology.

Autofocus is a tricky area. "You can do the research for 10 or 20 years," he says. "There's no limitation."

Traditional autofocus places a single item in the center, bounces an infrared or ultrasound beam off it, and judges distance by the speed of return. Once it knows the distance, it can focus the lens. A centered subject, however, often makes for a dead picture, and many photographers prefer to place the highlight to one side. Moreover, the viewfinder often shows two or more possible subjects at different distances, forcing the device to choose. Existing cameras were stymied.

When Canon first asked Kaneda to use fuzzy logic, he didn't think it would work. But other solutions had eluded him and, to his surprise, fuzzy logic succeeded. It allowed the camera to consider three possible targets—left, right, and center—and focus on the most likely. He placed a series of rules in the camera, such as "When the left is nearest, the likelihood that the subject is left is high." The camera could then assess how much the sensor readings satisfied the rules and choose the target.

Canon has also placed fuzzy logic in its copiers, to make better copies. The most important part of the copying process is the application of high voltage near the drum. Proper voltage is critical to picture quality, and three main factors determine it: room temperature, room humidity, and the density of the original picture. Canon engineers developed a fuzzy system that selects the best voltage given these conditions, and yields sharper copies.

Kato, who calls himself a "fuzzy evangelist," notes that some conventional engineers in Japan still oppose fuzziness. He feels their hostility partly reflects a sense of the need for models, which fuzzy logic economically eschews. "One of the major criticisms of fuzzy logic is that there's no design rule," he says. "In conventional technology, it's very important to make a design rule: how to design the system and how to make a model. In fuzzy logic, there's no rule to design or make a model, and that's probably why people don't like fuzzy logic. That's the most important point."

The attitude has changed at Canon itself. When starting any business development group, Sawada says, people now think of fuzziness early. Canon has also noticed that young engineers who have studied fuzziness at college are particularly enthusiastic. As the once-skeptical Kaneda says, "Fuzzy logic has gotten citizenship."

ON THE TUBE

By 1990, Sony was selling a fuzzy television in the United States that automatically improves contrast, brightness, color, and sharpness. "We were the first to bring fuzzy logic to TV products in this country," says training manager John Havens. Since these TVs optimize the picture, they also maintain stability across the channels. "One station may have a stronger signal level than another," Havens notes. "Ordinarily you might have to adjust it yourself. With fuzzy logic, the TV does it automatically." Sony's top-of-the-line Trinitron XBR checks the picture content every 1/60th of a second, monitoring 248 points of the video image. It then consults a database of 40 perfect images taken from a laserdisc and determines which aspects of the picture to adjust.

Viewers can evaluate the success of this feature for themselves, since they can not only turn it on and off, but alter it to match conditions in the room. If the lighting is low, for instance, they can call up the combination of brightness, contrast, and sharpness that best mitigates it.

THE BIRTH OF A BUZZWORD

Such products have made the term *fuzzy* a near-open sesame to consumers' wallets in Japan, and marketers have not missed this opportunity. Indeed, the parade of smart products has been a gift from the sky to them. "In Japan, they hear, 'Fuzzy, fuzzy, fuzzy,' all the time. It's a gimmick to sell things," says fuzzy scientist Elie Sanchez.

Puffery has afflicted the field. The fuzzy vacuum cleaner and rice cooker, for instance, are useful but not especially fuzzy. A number of consumer products use a diluted fuzziness, mere percentages or look-up tables rather than a fusion of rules. A few firms even paste *fuzzy* on utterly nonfuzzy products, and the nation is currently working toward better labeling standards. It can be hard to tell how fuzzy a "fuzzy" product is, but in general, the more finesse, the more fuzziness. The hype tends to arise with cruder items.

In a few cases, some U.S. researchers claim that better sensors have accounted for the performance Japanese marketers attribute to fuzzy logic. This charge too may be true. However, the same critics say it holds for one or two items from particular firms, not across the board.

What does the oversell imply for fuzzy logic itself? At worst, it means the watered-down form alone has created appealing advances. It has beaten conventional technology with one hand tied behind its back. In fact, Japanese engineers, who readily admit the marketing excesses, also express their confidence in the technology itself. Masayuki Oyagi, director of fuzzy projects at Omron, sums up the bedrock Japanese attitude: "What we think is important is the actual merits of fuzzy logic—not the name, but the convenience. As time goes on, more and more people will know the good things in fuzzy."

FUZZY EYES AND ELEVATORS

Such complaints do not affect the larger industrial sphere. Fuzzy logic began in industry, and after the second IFSA conference, it proliferated there also. Space forbids an exploration of them all, but a few examples will suggest the range.

In 1990, to the surprise of pedestrians, a large hole suddenly yawned open in the ground above excavations for a Tokyo subway line. Down below, a shield tunneling machine had removed too much earth and the surface gave way. Yasuhide Seno, an assistant manager of civil engineering at Okumura Corporation, says fuzzy control would have prevented this cave-in.

Shield tunnelers automatically scoop out rocks and dirt, place it on a conveyor belt, and carry it away. If subterranean earth were all the same consistency, the shield tunneler could remove it at an even pace and collapses would rarely occur. But it varies in density, so without intelligent control the tunneler can send back very different amounts, and if it excavates too much, the roof can fall in. Hence, skilled operators have had to run these machines. Okumura has developed a fuzzy controller to regulate this process, and it is now installed in several tunnels in Japan.

Everybody who works in a high office building knows the phenomenon. One spends several toe-tapping minutes waiting for an elevator, only to hear the ding-ding-ding of three arriving at once. Mitsubishi Heavy Industries has developing fuzzy elevators, which attack this annoyance. They help prevent convoys and cut the dead time in front of the elevator door.

Mitsubishi is the largest corporate entity in the world. It is the biggest of Japan's six major *keiretsu*, descendants of the old *zaibatsu*, and has holdings in banking, shipping, insurance, chemicals, oil, metal, real estate, and many other areas. Nikon and Kirin, for instance, are both Mitsubishi companies, and Mitsubishi Heavy Industries stands near the center of this vast web.

Tōyō Fukuda, a manager of advanced systems at Mitsubishi Heavy Industries, began working on the fuzzy elevator in 1986 and finished it in 1990. "We have to manage elevator movement to satisfy the passengers and they are calling us randomly," he says. Previously, he notes, Mitsubishi engineers "had many ideas how to manage the elevators, many, many ideas, but we had no way to implement them." He and his group thus turned to fuzzy logic, with its dextrous rendering of verbal rules.

For instance, experts say that if a call comes from a high story, the car should go there. But what is *high*? In conventional logic, the 10th floor may be high, the ninth floor not high. With fuzzy logic, of course, memberships describe *high*. "We have many other expressions which should be expressed by fuzzy membership functions," he says. Time of day is one. For instance, around noon, when people descend en masse for lunch, the elevators require a different pattern than at, say, 3:30 P.M.

In computer simulation, he says, fuzzy elevators slash waiting time by 50 percent, though he cautions that in-field figures are hard to come by, since one would need to compare a conventional and a fuzzy elevator in the same building. The new Kobe City Hall uses Mitsubishi fuzzy elevators, as do the 48-story Tokyo City Hall and several smaller buildings. Toshiba and Hitachi have also developed such elevators.

Mitsubishi is enhancing elevators in other ways. For instance, it has a display showing passengers how long they must wait for an elevator. It is shaped like a hourglass, and the last grains of sand trickle out of it just as the doors open. The company has also devised a way to measure crowd size at any one floor. A video camera monitors the waiting area and a computer evaluates the congestion from the image. The elevator can then give preference to floors where the most people are congregated.

Perhaps the ultimate vote of confidence for fuzzy logic will lie in controlling nuclear reactors. Because of Hiroshima and Nagasaki, nuclear power is a sensitive issue in Japan, even more so than in the United States. Nonetheless, Japan has few natural sources of power and it now obtains 27 percent of its total energy from 42 nuclear reactors. Mitsubishi Atomic Power has built 17 of them and is investigating the use of fuzzy logic to manage reactor cooling.

It began, as Japanese ideas often do, with a spur from below. In 1988 engineer Masahiko Hishida was impressed by the Sendai subway system and felt fuzzy logic might help cool nuclear reactors as well. A small group formed at Mitsubishi. "We've had many groups investigate how we can improve our company or our technology," says manager Tadakuni Hakata. "This group studied fuzzy logic and recommended that we study and apply it. Then my president is very fond of fuzzy logic, so he decided to push us to fuzzy." In 1989 the company retained its old Marxist labor agitator Michio Sugeno, who advised them on implementing it.

Cooling is vital to nuclear reactors. If the cooling system in a human being—perspiration, panting, dilation of subcutaneous blood vessels—fails, the individual dies of heatstroke. If it fails in a reactor, the core can melt down, a spectacular disaster. Hence, Mitsubishi must move slowly. "People in utilities are very conservative," Hakata says. "So I think they understood that fuzzy logic could do a better job, but we had to show them it's very reliable also." The company proceeded in steps, proving the dependability of fuzzy logic in one small control system, then another, and so forth. The evidence mounted, Hakata says, and soon Mitsubishi prepared to apply fuzzy logic to the cooling process itself.

"There may have been nervousness when we decided to adopt it initially, but now we're at the development stage, so we are not so nervous," says Hakata. Mitsubishi is simulating fuzzy control on a minisupercomputer and in 1992 planned a validation study on the simulator. The firm will also make fuzzy logic hardware and test the components.

In early 1990, Fujitsu announced a fuzzy electronic eye that distinguishes items as clearly as people can. It employs fuzzy logic to analyze scenes from a video camera. First, it estimates three-dimensional shapes from their two-dimensional outlines. Because it registers a new image every 30 seconds, it also detects the speed and direction of rapidly moving objects. Hence, it can predict where they might collide.

Fujitsu says it has successfully tested the fuzzy eye with driverless cars. The eye has automatically steered these vehicles around corners and away from obstacles. The company also plans to use it in robots that pick up objects on assembly lines, where, if successful, it could greatly streamline production.

THE HONDA PRIZE

As more and more Japanese grasped the virtues of fuzzy logic, they began nominating Lotfi Zadeh for awards. In July 1989, he was at a Tokyo hotel when he received a note informing him he had won the Honda Prize. The prize, initiated in 1977 to honor technology that fosters a "humane civilization," brought him ¥10,000,000 (about $77,000) and placed him in the company of Carl Sagan, Ilya Prigogine, and other well-known scholars. It was his first international award of any kind, and it surprised him, because hostility and skepticism still remained toward fuzzy logic.

"My experience is that, with respect to nominations, it's very easy to blackball a person," he says. "On a committee of 10, if one person says something negative, that's sufficient. Not just for prizes, but for nominations of any kind. People in general are not willing to argue. That's why if something is controversial, even to a slight degree, it's difficult to gain recognition for it."

He was not the only one surprised. When he formally accepted the award in November 1989, most engineers in the United States had never heard of fuzzy logic. Yet the prize bore the imprimatur of Honda, a solid corporation known all over the world. Somewhere, somebody was missing something.

9

TURF WAR

Anything that can be done with fuzzy logic . . . can better be done with probability.

—DENNIS LINDLEY

When the only tool you have is a hammer, everything begins to look like a nail.

—LOTFI ZADEH

■

One August evening in 1988, George Klir and Peter Cheeseman mounted the stage of a hallowed scientific debating hall at St. John's College, Cambridge University. Each spoke for one hour, then 20 minutes in rebuttal. Cheeseman argued that probability is the best way to deal with all forms of uncertainty, Klir that it isn't. An aisle halved the audience into supporters and opponents of the proposition. Rigid rules guided the affair, and a timekeeper made sure no one strayed beyond the limit. Afterwards, spectators voted with their feet, exiting through one of three doors: Cheeseman, Klir, and abstain. It was so formal, Klir thought, it was almost funny.

Klir ("kleer") is the sort of man one can spot across the room for his distinguished bearing: upright stance, bristly gray beard, merry eyes. That night, he gazed out at an audience of probabilists, opponents of fuzziness. He found several a little arrogant, but many hearty and good-spirited. Some even sallied forth to defend him in his near-isolation. But the most remarkable thing, he felt, was the poll itself. A majority abstained, and among the 50 or so voting, Cheeseman won by a hair. Cheeseman later joked that everyone had drunk a lot of wine and that Klir's exit led to the men's room.

Yet for all its cheer, the debate reflected a serious tide of the late 20th century: the rise of uncertainty and the resulting battle among theories of it.

Uncertainty is the stuff of life, the world we live in. It is the subtle smile, the half-sensed pattern, the vexing decision. It is tomorrow's stock market, the best chess move, the baffling ailment. It is so ubiquitous that scholars for centuries have studied how to escape it rather than how to work with it. They have preferred the solace of certainty to a grasp of uncertainty.

Yet computers have brought uncertainty to the fore. These machines take us deeper into complex systems, and as a result we want to study even more complex ones. "When the complexity grows, we have to give up something," says Klir. "One thing you can give up is the certainty."

How do scientists deal with uncertainty? This question has spawned a host of constructs and a tournament for territory, of which the

Klir–Cheeseman debate was one joust. For the participants, the stakes are huge. They involve careers, grants, prestige, influence, and fame. Should one theory prevail, it could proliferate in expert systems and other forms of machine intelligence, and its supporters could catch a high wind to the future. The others could enter the doldrums.

Fuzzy logic was not first in this arena and hence faced a problem of priority. After settling the terrain of vagueness, it found many hand-driven signs claiming the land for Bayesian probability. "As fuzzy became somewhat more popular the intensity of their attacks increased," Zadeh says. "Previously they felt it was something that was not worth talking about. It was a minor irritant. But as more people accepted it and applied for grants, as it became more visible, it became more of a threat."

Bayesian opposition was a major reason fuzzy logic languished in the United States in the 1980s. American corporate engineers were unfamiliar with fuzzy logic and ignorant of advances in Japan. They relied on experts like the Bayesians, who dismissed fuzzy logic, claimed universal superiority, and, Zadeh says, scotched many fuzzy research proposals. Directly or indirectly, these scholars deterred many businesses from probing fuzzy.

As the fuzzy products from Japan became better known, their salvos continued. Eugene Frankel had said that a new paradigm must not only solve problems, but win credit for solving them. Hardline Bayesians deny fuzzy logic this credit. For instance, what do they think of the Sendai subway?

Judea Pearl, a professor of computer science at UCLA, says, "If you want to make a smooth-ride system, I can make you a smooth-ride system. I just design a smoother controller. . . . I'm wondering if [fuzzy logic] contributes anything to what they would have been using otherwise."

"My understanding from talking to control engineers," says Cheeseman, "is that it is such a simple problem that they could do it with perfect ease."

One control engineer is Werner Remmele, head of intelligent control systems at Siemens, the huge German electronics firm. Siemens makes subway controllers, and Remmele, who has carefully studied the Japanese competitor, snorts at the Bayesian claims. "Do you know one system that runs as smooth as Sendai?" he says. "Do you have one worldwide? There is none."

Michael Reinfrank, also a control engineer, manages the Siemens lab on intelligent control systems. "Fuzzy logic represents the expert information from the subway driver," he says. "It's next to impossible to represent this in any other form except fuzzy."

Yet extreme Bayesians remain adamant. For centuries probability has been the only tool for handling uncertainty, and they insist it still is. Anything fuzzy logic can do, they say, Bayesian probability can do better and cheaper.

So why have the Japanese embraced fuzzy logic? "It's a fad," says Cheeseman. "Basically it seems to be promising something for nothing is what is going on here."

Who are these people?

EFFECT AND CAUSE

Dismissed as a small and trivial sect not long ago, Bayesians are a major, rising force. They have won impressive successes, and they occupy chairs at prestigious universities, referee grant applications and articles, and stage lively debates. They know their field has flared and guttered in the past, and to prevent its dimming again, they speak clearly, forcefully, and even eloquently.

They see probability as a dominant part of reasoning, crucial to thought and survival. Judea Pearl depicts the brain as alive with a subtle network of likelihoods. "The human mind is so rich with frequencies, albeit invisible ones, that only a calculus tailored after frequencies can manage the outcomes of this experience," he once wrote. "These invisible frequencies involve track records of mental performance, past usage of concepts and strategies, and vast populations of synthetic scenarios. Perhaps even the firing activities of neurons are relevant here."[1] Such processes lie at the core of the mind and probability animates them all.

And Bayesian probability in particular.

The better-known kind, called *objective probability*, arose around 1660 in response to queries from European gaming houses. It is simple. Past frequencies make good predictors. If a coin has come up heads half the time in the past, it will come up heads half the time in the future. In other words, the coin has a 50 percent chance of heads on the next flip. Relative frequency can light the way before us, and because they rely on it, objective probabilists are sometimes called *frequentists*.

At this level, objective probability moves from cause to effect. Flip a dime (cause) and there is a 50 percent chance of heads (effect). It starts with hard facts that are easy to test, hence difficult to dispute. Not only does it plainly work—witness its neon monuments in Las Vegas—but it removes the taint of personal opinion, a trait appealing to scientists.

Bayesian probability is quite different. It began with a man who never published a mathematical work in his life and who might be surprised to hear of the feisty discipline that has sprung up in his name. Thomas Bayes (c. 1701–1761) was a Nonconformist minister in Tunbridge Wells, some 35 miles southeast of London. Despite his obscurity, he must have shown talent, since his contemporaries elected him a Fellow of the Royal Society.

After his death, his friend Richard Price discovered "An Essay toward solving a Problem in the Doctrine of Chances" among his papers. Price read it to the Royal Society in 1763 and had it published in the Society's influential *Philosophical Transactions* in 1764, adding an introduction and an appendix. His contemporaries ignored it, perhaps partly because it made uphill reading.

But it contained Bayes's Theorem, the keystone of Bayesian probability. This formula explains how to update a probability when new evidence arrives. The classic example involves an urn filled with black and white balls. Suppose an individual believes the ratio of black to white balls inside is 50/50. However, as she begins extracting them, she gets three black balls for every white one. This evidence may spur her to change her expectations, and Bayes's Theorem shows how to do it.

This kind of inference is called *inverse probability*, reasoning back from effect to cause, or more precisely, from evidence to hypothesis. Objective probability asks, "What is the chance the next ball is white?" and bases the answer on the ratio of balls in the urn. Bayesian probability asks, "What is the ratio of balls in the urn?"

It remained a curiosity until Pierre Laplace (1749–1827) came across it. This French mathematician, who inhabits the same celestial circle as Newton, Euler, and Gauss, essentially founded Bayesian inference. Bayes's Theorem revises probabilities but does not reveal how to form them in the first place. What if we have no idea of the initial ratio of balls in the urn? We have nothing to update. Laplace declared that, in total ignorance, we should allot probabilities equally among alternatives. According to this controversial *principle of indifference*, if the urn is terra incognita, we assume a 50/50 chance of white or black before drawing the first ball.

Laplace took this stance because he viewed probability not as relative frequency, but as expectation, as degree of reasonable belief. Since probability is expectation and in utter ignorance we have equal expectations of black and white, we should assign them equal probabilities.

He thereby introduced a crux of Bayesian probability: subjectivism. Laplace guessed at values when necessary. For instance, he did not hesitate to suggest that a witness might be 60 percent likely to tell the truth. He was hardly the first to take this shortcut, but, like fuzzy logic, it seemed to mark a departure from the rigor of science.

However, the Bayesians argue that they don't deduce probabilities, but estimate them. They aren't defying relative frequency, but venturing beyond its confines, and they claim they can bring back results that make it worth the trip. They also say most people can guess simple probabilities fairly accurately.

More broadly, they allege that probability is inherently subjective. All of it measures degree of belief. Even the hard numbers of objective probability—the chance of a royal flush, an airplane crash, a winning lottery ticket—express our reasonable belief that these events will happen. It is misleading to say heads has a 50 percent chance of occurring. Rather, we have a 50 percent belief it will occur. Probability calibrates anticipation. It can be no other way. Certainty is simply total belief, and uncertainty, partial belief. Fractional expectation is the soul of all uncertainty.

Overall, Pearl says, three flags signal Bayesian probability:[2]

1. It reasons backward from evidence to hypothesis.
2. It accepts subjective evaluations.
3. It builds a total model of the situation, creating an edifice in which all probabilities add up to 100 percent, rather than assessing individual frequencies in isolation.

After Laplace, other well-known thinkers took the baton, including logician Augustus De Morgan (1806–1871) ("Probability is a feeling of the mind, not the inherent property of a set of circumstances"[3]), astronomer and natural historian John Herschel (1792–1871), mathematician Siméon-Denis Poisson (1781–1840), and economist William Stanley Jevons (1835–1882).

Beginning around 1840, however, subjectivism met harsh criticism from objectivists like John Stuart Mill, George Boole, John Venn (1834–1923), and others. Their views prevailed, and probability became oriented to large samples. Thus commenced the long dark night of subjective probability. According to Bayesian statistician E. T. Jaynes, this eclipse was "the most disastrous error of judgment ever made in science."[4]

Yet objective probability revolutionized science. In the 1870s, James Clerk Maxwell (1831–1879) and Ludwig Boltzmann (1844–1906) took a hint from insurance tables and applied statistics to the behavior of gases, discovering the laws of thermodynamics. Gustav Fechner (1801–1887) introduced probability into experimental psychology. In 1897 Emile Durkheim (1858–1917) published his famous statistical analysis of suicide, showing large-scale regularities in this impulsive act. Max Planck (1858–1947) started from probability en route to quantum mechanics. And the heredity theory of Gregor Mendel, with its dominant and recessive traits, was intrinsically probabilistic.

From around 1890 to 1930, statisticians devised the techniques that now reign in the social sciences: the correlation of Francis Galton (1822–1911); the correlation coefficient, multiple regression, and chi square of Karl Pearson (1857–1936); the t-test of William Gosset (1876–1937); the analysis of variance, or ANOVA, of Ronald Fisher (1890–1962); and the confidence intervals of Jerzy Neyman (1894–1981). Some, like Pearson and Fisher, embraced statistics to further their dream of eugenics, but their work brought probability beyond the stage of exposing regularities in mass phenomena and enabled it to show causation.

However, subjectivism never entirely succumbed, and British geophysicist Harold Jeffreys (1891–1990) began its revival in 1931. This lifelong foe of plate tectonics was having trouble analyzing earthquake data and began probing statistics itself. The result was his controversial book *Scientific Inference*, which refined the rules of Laplace and sparked a running debate with Ronald Fisher.

In succeeding years, Leonard Savage laid down axioms for Bayesian probability, and Bruno de Finetti, E. T. Jaynes, and others developed it further. Partly because of their work, Bayesian inference grew rapidly in the 1970s and especially the 1980s. Cheeseman attributes this rise partly to the advent of cheap computing. "Anyone in a lab can now have a pc on their desk and they can actually do online complex Bayesian inference, numerically," he says. "Before it had to be done by hand with a calculator and it was just unfeasible."

Yet the debate between objectivists and Bayesians continues, and compared to it, the fight with fuzzy logic remains a skirmish. Most statisticians, and virtually all scientists, remain objectivists. Why? Stanford professor Bradley Efron set forth a few of the reasons in 1986.[5] For one thing, he said, objective probability focuses more closely on the foundation, the raw data. For another, it solves subproblems independently, while Bayesian probability must erect the whole edifice. Moreover, scientists must assure themselves and others that their results are fair, and objective tools have much more appeal in this regard. This kind of analysis ignites reaction. Ultra-Bayesian Dennis Lindley called it "an attack on a parody of a serious argument" and added, "Every statistician would be a Bayesian if he took the trouble to read the literature thoroughly and was honest enough to admit that he might have been wrong."[6] It is a strong statement: Non-Bayesians are lazy, pigheaded, or both.

Indeed, a heat-shimmer of controversy surrounds Bayesian probability. "One thing that amazes me in this field is that people hold these various viewpoints with near-religious fervor," says Cheeseman. "There are almost standup fights at the congresses over the question of how you assess probabilities." Or, as Bayesian E. T. Jaynes characterizes it, "It's a jungle out there, full of predators tensed like coiled springs, ready and eager to pounce upon every opportunity to misrepresent your meaning, your purpose, and even your results."[7] The Bayesians themselves are no lambs in this jungle. "The soldiers of the Bayesian church are a particularly quarrelsome lot," says John Fox, "with a well-known reputation for ruthlessness in pressing their Belief."[8]

The more hard-line Bayesians maintain that probability is the only legitimate way to handle uncertainty. Even words, they say, yield easily to probabilistic analysis.

Judea Pearl is a likeable man with a winning smile and an expansive manner. He gestures as he speaks, and when he smiles in making a point, his eyes narrow till they twinkle. Born in Bene Beraq, just northeast of Tel Aviv, this major figure in the Bayesian world earned degrees in the United States and worked nine years on electronic memories before becoming at professor at UCLA in 1970.

Though he claims ignorance of much of fuzzy logic, it does not impress him. Words are its supposed stronghold, he says, yet probability rules here as well. If there is a half-eaten apple in the refrigerator, fuzzy theorists say there is half an apple. Pearl would ask the probability that it would "be proclaimed an apple by a person in our culture under the circumstances." More generally, he says, "The question to ask is what is the probability that a person having an intention I would utter the word W to express his intention." That is, one asks the likelihood that a person intending to describe the half-apple would use *apple*. If the chance is 50 percent, then the item is an apple to 0.5 extent.

Max Black took this approach and its pitfall is evident. People round off. An individual might think the Mississippi is 0.8 *long*, yet always call it *long*. Had Eleanor Rosch asked *whether* chickens, ostriches, and robins were birds, she might have gotten uniform yeses. Instead, she asked *how much* they were birds and found a sliding scale. When Willett Kempton forced his subjects to give yes/no responses, they said *mug* was a total subset of *cup*. When allowed to state degrees, they said *mug* was a partial subset of *cup*. Rounding off hides information.

Bayesians and all probabilists work within the confines of bivalence. Hence, they must recast *how much* as *is or isn't*, and "How tall is Tom?" becomes "Is Tom tall?" asked of a thousand people. But it does not quite work, since probability and fuzziness describe different facets of words. Probability measures when people use them, fuzziness how much they mean them. Both are useful, but they are not interchangeable.

Pearl has further criticisms. He likens fuzzy devices to "ad hoc engineering," solving problems on the fly. "I'm thinking to myself, if I were to design a washing machine now, I would probably do what the Japanese do right now, and believe me, I know nothing about fuzzy logic," he says. "Somebody would tell me, 'Make sure the machine scrubs faster when things get dirtier,' and I would translate this into some relations between two quantities, and I'd probably go through the same process the engineers are going through now in Japan, without calling it *fuzzy*. Just fine-tuning of ad hoc engineering practices, which are workable."

Like other Bayesians, he makes an additional claim: By forsaking probability, we surrender a rich tradition of theory and experience. Its roots go back centuries, while those of every other theory are decades deep, at most. "Probability offers you well-defined constraints about the type of manipulations which are permissible on sentences," Pearl says. "They are well defined, they are understood, and in my experience they never lead to counterintuitive conclusions." But fuzzy logic, he says, rips out and flings away these guardrails, and he would like to see some salutary thou-shalt-nots.

He adds that probability is easy. "The functions are clear," he says, "therefore the program you write is easy to debug, easy to tune, easy to refine, easy to enrich, and easy to extend. Because you know how one portion of the program should affect another." He does not say the program is easy to write in the first place, and this issue is a matter of dispute.

CATS, MEN, AND STARS

"I've actually tried to pin Lotfi down on this: What is it that fuzzy reasoning is doing that probability reasoning is not?" says Peter Cheeseman. "The best answer I've got out of him is that fuzzy logic is easier to use. And I think in the day of computers that is just an irrelevant consideration.

"They can turn around and point at controlling a train. We can point at controlling nuclear submarines and shuttles and going to the moon," he says, referring to the Kalman filter. "The best probabilistic control theory trivializes the other applications beyond comparison. And I would just say the reason [these consumer products] haven't been done with probability is that it's just so simple, no one ever thought of bothering."

Cheeseman grew up in Australia, where as a youth he fired six-foot rockets 30 miles into the air. "I'm very lucky to be here today," he laughs. He came to the United States in 1980 and began working at the Stanford Research Institute (SRI) on PROSPECTOR, an early expert system. For the last five years he has been in the AI group at NASA Ames, where he has developed a Bayesian program that discovers classes in data. Says fuzzy theorist Hamid Berenji, who also works there, "We always argue with him. That's fine. He's a good friend of mine."

Around 1982, Cheeseman invented his own uncertainty calculus, a kind of averaging technique. "I was very proud of this," he says. "I felt sure this was how to do uncertainty reasoning in expert systems." To show its power, he set it head-to-head against Bayesian inference, which triumphed utterly. "And at that point I became a convert. Then, when I understood the basic theory, I realized that's the only way to do the inference. That anything else has to be inconsistent."

He has read a great deal about Bayes himself. "I've even visited the tomb of the Reverend Bayes," he chuckles "I've got the original Bayes paper in the Royal Society proceedings. It's a facsimile copy but it's sitting proudly on my bookshelf."

He first heard of fuzzy logic in a curious way. One day at SRI he was chatting to visitors about Bayesian inference and one of them replied with talk of fuzzy sets. "It turns out it was Lotfi Zadeh. He used to visit SRI quite a lot. And he's a great guy."

Later they had long discussions and panel debates. Cheeseman is a skilled debater, but he feels uneasy scoring points at the lectern. "In science the goal isn't to defeat the opponent. It's to get at whatever the truth is," he says. "And trying to present just one side—I find that a very difficult thing to do." Yet his strongly pro-Bayesian case concedes little to fuzzy theorists.

First, he says Bayesian inference can perform every task fuzzy logic can, and therefore fuzzy logic is unnecessary. Why? "Probability is the calculus of belief and so says all that can be said about uncertainty."[9] Moreover, it is not only a richer system, but one more firmly grounded in fact. "The main difference is that, with fuzzy sets, you don't do explicit conditioning on the evidence [i.e., updating of probabilities], so the answer is not dependent on what evidence you use."

He also says fuzziness fails to reflect context. In "Mary is young," the term *young* has one range of meanings if we are discussing grade schoolers, another if we are talking about retirees. In Bayesian analysis, one evaluates how "Mary is young" changes the listener's expectation, which in turn depends on context, on prior probabilities.[10]

Fuzziness allows people with no engineering background to perform control. Is that good? "It's not bad," he admits. "If it works, it works, and that's the end of the story. If those people were trained in probability theory, they might get better performance."

Cheeseman agrees that some classes have no sharp boundaries, even if we possess all knowledge about them. "What is a cat? What is a man? What is a star? None of these things have very clear definitions," he says. But he parts with the fuzzy theorists on how to depict such categories. Like Pearl, he looks to the probability that a person would describe them as cats, men, and stars.

In 1987, Dennis V. Lindley, a retired professor from University College, London, reissued the Bayesian claim that "the only satisfactory description of uncertainty is probability."[11] All other means, including fuzzy logic, were "unnecessary."

Zadeh had said that the Bayes rule was inadequate since most information we deal with is fuzzy. Lindley replied that probability could easily replace fuzziness. For instance, he said, Zadeh alleged that the sentence "Berkeley's population is over 100,000" is fuzzy, because it implies that the population is not much over 100,000, and *not much over* is fuzzy. But what does that tell us? It is a rough, personal estimate. Why not base such values on the real world, Lindley asked, on propositions we can test? For instance, why not put the question to an informed Berkeley city official? She might fix the population at 115,000. We could then transform the answer into a probability given her knowledge.

He also asserted that fuzzy logic was more complicated than probability. "Fuzzy logic leads to nonlinear programming and contains great complexities of language and ideas. Yet probability is extremely simple, using only three rules and containing rich concepts like independence and expectation. . . . Is it not best to accept the advice of William of Ockham and not multiply entities beyond necessity?"[12]

In the year the Sendai subway debuted, and apparently ignorant of the fuzzy kiln, backgammon program, water purification plant, and expert systems, Lindley concluded, "The probabilities . . . , and only these, are the quantities needed for coherent decision making by a single decision maker. How can one use fuzzy logic . . . to decide?"[13]

THE TAXONOMY OF UNCERTAINTY

In 1964, George Klir was teaching electrical engineering and computer science in Prague. Out of the blue, the Czech government offered him a two-year appointment at a university in Mosul, Iraq. He hastily accepted, for intellectual life in his homeland was sterile. "The country was closed to information," he noted, "and one was always exposed to the danger of committing unwittingly a 'political crime.'" Mosul differed utterly from Prague, but for Klir, the highlight was access to knowledge. He could buy Western publications at newsstands and order almost any book he wanted through the mail.

When his term in Mosul ended in the spring of 1966, Klir and his wife decided not to return to Prague. They went instead to Vienna, applied for immigration to the United States, and waited months in bureaucratic limbo. Finally, in November, their visas arrived. They flew to Los Angeles, where he became a professor in computer science at UCLA.

At first, he says, he felt like Alice in Wonderland. He spent endless hours roaming the libraries and bookstores, savoring the vast tracts of literature on almost every topic imaginable. In fact, he soon realized it was overwhelming. He had to be selective. "Time has been my most precious commodity since our arrival in America; it is always in short supply."

As a student, he had read Zadeh's work on system theory and found it fascinating. In 1969, he published a text on systems science, but in general found little enthusiasm for systems ideas at UCLA. That year he joined the School of Advanced Technology at SUNY Binghamton, where he remains today. He is editor of the *International Journal of General Systems*, which he founded in 1974, and coauthor with Tina Folger of *Fuzzy Sets, Uncertainty, and Information* (1988), a major fuzzy text.

Klir notes that one problem with the Bayesian claim of across-the-board superiority lies in the narrowness of probability. Several theories are broader than probability and subsume it. One is Sugeno's fuzzy measures, and another is Dempster–Shafer theory, created in two stages by A. P. Dempster in the 1960s and Glenn Shafer in the 1970s. Dempster–Shafer theory adds a third wedge to probability: plain shoulder-shrugging uncertainty. Probability carves all possible outcomes into two abutting domains, so if an event is 40 percent likely, it must be 60 percent unlikely. But with Dempster–Shafer, if it is 40 percent likely, it might be 45 percent unlikely and 15 percent uncertain (see Figure 12). The theory weighs pro, con, and not sure.

Like most fuzzy scholars, Klir believes Bayesian probability can be very useful, but it is not a panacea. One doesn't take antibiotics for high

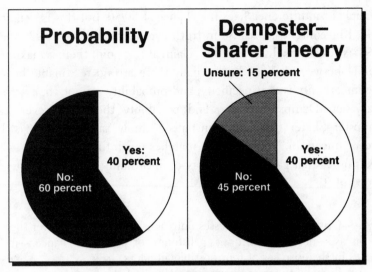

FIGURE 12

blood pressure. Probability, he says, addresses only one kind of uncertainty. There are at least four basic types: nonspecificity, fuzziness, dissonance, and confusion.

Nonspecificity is basic ambiguity, a one-to-many relation between statement and possible meanings. For instance, a test might show a patient has either hepatitis, cirrhosis, gallstones, or pancreatic cancer. As among these four, it says nothing. It is nonspecific. Nonspecificity is lack of informativeness.

Fuzziness, of course, is vagueness, the degree to which a term like *cirrhosis* applies.

Dissonance is pure conflict. One statement is true and its rivals are false. For instance, the patient may have a liver ailment or a non-liver ailment. Some evidence supports one thesis, different evidence the other, and we are uncertain between the two.

Confusion is pure and potential conflict. Suppose one test tells if a patient has liver disease and a second if he has stomach disease, but that for some ailments both tests are positive. In this case, a positive liver test may indicate stomach disease. Not only is there conflict, but the very meaning of the evidence is unclear.

Probability cannot handle all these species of uncertainty, Klir says. For instance, suppose we pick a ball from an urn, knowing nothing of the ratio of black balls to white. Will it be black or white? Laplace's principle of indifference fixes the chance at 50/50. But this is sleight of hand, Klir contends, because in truth every probability—78/22, 55/45, 1/99—is equally likely. Probabilists claim at this point there is maximum conflict between beliefs, but actually there are no reasonable beliefs. The uncertainty is nonspecificity, and probability cannot cope with it. Fuzzy set, Dempster–Shafer, and possibility theory, however, can. Dempster–Shafer theory, for instance, simply says the case is 100 percent uncertain. This assertion matches our intuition.

Klir notes that each major theory of uncertainty deals with different sectors of it:

- Crisp set theory addresses only nonspecificity. Whatever the disease—hepatitis, cirrhosis, gallstones, or pancreatic cancer—it lies in the crisp set of four possible ailments.
- Probability treats dissonance. Which of the four is it?
- Fuzzy set theory handles nonspecificity and fuzziness. To what extent is it cirrhosis?
- Possibility theory works with nonspecificity and confusion. How possible is it that the positive liver test indicates stomach disease? Possibility theory can measure confusion because it assesses how easily an event can happen, its potential.
- Dempster–Shafer theory treats nonspecificity, dissonance, and confusion.

Probability handles only dissonance. It fumbles the other three. Klir says this "undeniable fact" voids the claim that probability is the only way to describe uncertainty.[14]

Klir also criticizes assigning a precise number to the degree of belief in any proposition. It is simply unrealistic, he says, to demand a 75 percent belief, say, when our level of expectation is usually hazy. It forces us to be certain about our uncertainty.

Moreover, fuzzy logic has won clear successes in control applications, and he notes, "It is not clear to me how probability theory could effectively describe and manipulate the great variety of descriptions or rules that are possible in natural language."

He concludes that probability models only one kind of uncertainty—degree of conflict—because it was designed that way. He suggests that those who extol its perfect superiority heed a verse in the *Dao De Jing*:

Knowing ignorance is strength.
Ignoring knowledge is sickness.

By making vague expectations precise, he says, the Bayesians fail to know their ignorance, and by disdaining to model vagueness appropriately, they ignore available knowledge.

Suppose Klir is wrong and Bayesian probability can describe all uncertainty. Is it always the best method? Glenn Shafer, of the University of Kansas, says the holder of this view "is no less correct than the man who believes that the only satisfactory household is one with a dozen servants. It's wonderful if you can afford it."[15]

In fact, many observers cite the impracticalities of Bayesian probability. John Lemmer states, "There are known serious hazards on the sea of Bayesian knowledge engineering."[16] Stephen Watson observes that probability theory leads "to enormous problems of complexity, and as a matter of practice it is necessary to seek for approximations."[17] Fuzzy theory is a useful tool for handling imprecision, he adds, since it simplifies the problem.

A. P. Dempster and Augustine Kong, of Harvard University, note that though Bayesian decision analysis has been in the field for some 30 years, it has penetrated its natural markets to only limited extent. If potential users were simply ignorant of it, they say, then Bayesian polemics might have a point. But "a more plausible explanation is that practical construction of realistic Bayesian models is typically very difficult."[18] They too agree that fuzzy logic has a proper place in formal analyses.

In other words, in the real world Occam's razor cuts against Lindley.

So do other concerns. The very sweep of the Bayesian charge—*always*, not *usually*—invites skepticism. "Notwithstanding claims that there is only One True Way of conceiving uncertainty," says John Fox, "a consideration of the history of physics, philosophy, politics—and AI—will reveal that there is rarely one true representation of anything."[19] He could have added linguistics, perception, anthropology, psychology, astronomy, and much, much more.

Moreover, the blanket nature of the claim implies censorship, suppression. If scientists accept it, they will cease exploring other avenues, which must be fruitless. The realm of inquiry will become closed and impermeable rather than open and self-correcting. If Cheeseman and Lindley are wrong, we will not find out for a long time. We could therefore suffer dramatically in consumer products, expert systems, and many other fields. "Do not put all your fortune in hollow ships," said the Greek poet Hesiod,[20] and stockbrokers likewise advise spreading investments out in a portfolio to avoid disaster. Bayesians, however, urge that we place all on a single bet.

And they do so despite the tide of Japanese products. While Bayesian techniques have certainly succeeded in complex high-tech instruments, the consumer market reveals a cornucopia of fuzzy products and no Bayesian products of any kind. They simply don't exist. Perhaps Bayesian methods could outshine fuzzy logic here, but cost and ease of manufacture are paramount in this realm, and the Bayesians have yet to show any consumer triumphs at all.

The Bayesians admit of no exceptions to their claim. It is a crisp *always*, not a fuzzy one, so it falls apart with one counterexample. Go out to the end of a limb and the very twigs must support you. Yet broken twigs abound.

For instance, Kamran Parsaye, who studied under Judea Pearl, sells a software program that uses both Bayesian probability and fuzzy logic. "I don't think one can do the kind of practical things with Bayesian probability that one can do with fuzzy logic," he says. "There are too many mathematical constraints around the Bayesian probabilities. All the numbers have to add up to 100, and so forth. It is a far more complicated mathematical machinery." In developing his software, he says, he began with Bayesian probability, but at some point had to shift over to using fuzzy logic. "It was the only way we could get the program to work."

Notified of Parsaye's program, Cheeseman said he hadn't heard of it, but that in "every case I've seen where people say you can't do something with probability, you can and it's very straightforward."

Cheeseman dismissed Yamakawa's inverted pendulum as "the most elementary control problem imaginable. When you get to a double pole, it's a very hard problem, because you get down to two angles." Informed that Xiwen Ma in Silicon Valley had built a fuzzy system that balanced three poles atop each other, an unprecedented feat, Cheeseman was surprised. "I've only seen the one," he said. "Certainly balancing multiple poles is a very hard control problem." So did the triple pendulum impress him? "Not really," he said, "because these are the kinds of things that are student exercises in control theory. I don't know about the triple pendulum, but certainly the double pendulum is. So the fact that this approach is able to get something that students achieve as a regular part of their course just doesn't impress me."

In fact, the relation between fuzzy systems and probability held further surprises for everyone, as a young professor at the University of Southern California (USC) was about to show.

10

AMERICAN SAMURAI

■

My theorems are false if and only
if 2=3. So you don't have to like
it. You just have to take it.

—BART KOSKO

The fact is that what Einstein did
with gravity was to eliminate it.
That's what we do here: eliminate probability.

—BART KOSKO

In 1985 Bart Kosko sat in the packed hall at an AI conference at UCLA. On stage were Peter Cheeseman, Lotfi Zadeh, Judea Pearl, and other luminaries. With the flair of a standup comic, Cheeseman gave a once-over of Bayesian probability, lightly taunted Zadeh, and issued a challenge: $50 to anyone who could solve any problem Bayesian inference could not. By then he had completely won the crowd over. Zadeh rose to speak soon after, and Kosko watched in disbelief as he ignored the blazing gauntlet. Audience enthusiasm turned to stone.

"I really felt a cold chill in my spine," Kosko says. "It was the first time that I fundamentally thought fuzzy logic was a tremendous mistake. I really thought the old vessel simply couldn't hold water. And when push came to shove, when you finally got out of the fuzzy ghetto, you got pushed right back into it. Even the Grand Master couldn't hold his own against an upstart."

At that moment he decided to plumb fuzzy logic to its depths, to determine once and for all if it had substance. If it existed, he would pursue it out to the end. If not, he would flay it so savagely that he'd "make Cheeseman look like nothing."

Kosko is a singular individual, brilliant, brash, self-disciplined, competitive, and highly controversial, even within the fuzzy community. He is a 33-year-old polymath at USC, a mathematician, engineer, black-belt karate expert, screenwriter, novelist, bodybuilder, and composer of symphonies and sonatas. In his spare time, he scuba dives, shoots trap, and hunts wild boar with bow-and-arrow. His sci-fi novella *Wake Me When I Die* has two morals: "No pain, no brain" and "Comfort kills."

He is a Coleridgean talker, pouring forth ideas so rapidly that students compare listening to his lectures to trying to drink from a firehose. The references are wide-ranging: Ashoka, Hasidism, the Coase theorem, the murder of philosopher Moritz Schlick, the literary theories of Mickey Spillane. Zadeh describes him as "intense, very intense." He says, "I'm a fuzzy positivist and I have to prove myself."

He threw himself utterly into fuzziness and not only decided it was real, but developed a new and all-embracing model of the field. Its base, which he calls *subsethood*, led him to many of the root concepts of math in general and probability in particular.

"Zadeh is a much nicer guy than I am," Kosko says, leaning back in his chair. "He won't argue with other people's rules. But that doesn't convince the mathematicians. You have to step into the ring, wear their boxing gloves, beat them at their own game. And that's where I come in.

"I have results I can prove and I throw an open challenge: I'll fight anyone, on any conditions, any terms, provided it's convenient. I have theorems and I have proofs. You've just got to take it on the chin. There's nothing you can do about it. I didn't walk into battle without a sword on."

He was born in 1960 in Kansas City, Kansas. "Bart was known by the other kids for his passion," says Kevin Helliker, a *Wall Street Journal* reporter who grew up with him. "Bart didn't *sort of* like anything. If he liked a movie, he might see it 20 times. If he didn't like a dish, he wouldn't eat it, no matter what the threat."

At 10, he ran with a street gang. "There were railroad tracks and we kind of controlled that. We were always fighting with other groups." His junior high repeatedly ejected him for schoolyard battles or excessively long hair. He also imbibed a wide array of drugs, but in 1974, just after junior high graduation, a bad trip left him unable to sleep and gave him headaches for days. As a result, he forswore drugs and rebelled against hippiedom itself, especially rock.

He was already interested in Bach and now committed himself to classical music. "I was really into it," he remembers. He quickly mastered the mandolin and other string instruments. A friend of his had an upright piano in his basement, and in September 1974, he learned to play it. By the end of that month, he was composing his first violin concerto. He was 14.

Soon he had written a quartet, an orchestral suite, and other works. He often labored far into the night, falling asleep at 3 or 4 A.M. and waking at 6 A.M. for school, 10 miles away. After two hours of sleep, he would feel ill in the morning and drag himself through the day, or simply cut class altogether. He missed school 50 times that year to write music. "You're at your best once a day," he says. "I never gave my best to school or homework."

By next fall, when he was a junior, musicians at St. Mary's College were rehearsing his first quartet, and soon the Kansas City Philharmonic played his first trio. Over the following summer he wrote 100 pieces, including his *Overture to the Count of Monte Cristo*, which he completed in two weeks. He then won a young composer's contest, which led to scholarships and placed his name in the paper. "Now I'm in my senior year, and to the great surprise of my friends and nonfriends, they open up the paper and here's Bart Kosko involved in music of all things," he says. "That's not how I was known in the school."

In December, he began applying to college. He was determined to go West and write, score, and direct movies, and USC gave him a full music scholarship. He arrived at USC and discovered that atonalists— who based their art on the 12-tone scale of Arnold Schönberg— controlled the department. He hated atonal music. He didn't even want to listen to it, much less write it, and he created an uproar. Ultimately, the faculty told him he could keep his music scholarship for one year if he left the department, and he switched to a dual-degree program in economics and philosophy.

Since USC was the third most expensive school in the nation, he was always angling for money, and soon he began writing pseudonymous fiction and articles for libertarian journals. "I had a work-study job at night, I was taking 18–20 units, it didn't matter, I was writing my political stuff, a novella here and there, just applying myself in an Aristotelian sense of happiness."

At 20 he made another fierce commitment: to master mathematics. "I knew I had to have it, just had to do it." He bought a calculus book. On his first pass through, he did not care for it. So he went back to basic high school algebra, factoring polynomials. By now he had a strong foundation in the philosophy of math and logic, and he quickly built on it.

In his senior year he walked into the class on differential geometry. "It's very hard stuff, but you cannot understand relativity without it," he says. The teacher was Mark Kac ("kots"), an eminent probability theorist, and he quickly became Kac's protégé. Kac taught him the importance of rigorous proofs and deflected his career once again, convincing him to take his master's in mathematics.

After graduating from USC in 1982, Kosko earned his master's at UC San Diego in one year. He then took a job as an internal consultant with General Dynamics at $30,000 per year, a sum whose magnificence he could hardly believe. The firm gave him a problem in allocating weaponry in case of attack. He soon checked out between 200 and 300 books from the corporate library, and so many articles lay scattered on his desk that General Dynamics changed its rules on photocopying. Amid this sprawl he discovered a paper by Ron Yager on fuzzy sets and decision theory. It referred to an intriguing 1965 article by Lotfi Zadeh. "And I read that and really felt: Here it was. I felt Zadeh had done something great." The world was imprecise and math was precise, and Zadeh had made them fit.

He began working on fuzziness, and proved a theorem fixing the size of fuzzy sets. In 1984 he saw Zadeh at an AI conference and approached him at the end of a day. "I have a new proof of the sigma count," he declared, and commenced a high-speed lecture, summarizing the proof "as fast as I could put it out. I was huffing and puffing. And he said, 'Very good. Write this down and send it to me.' And he waved me off.

"Well, I was excited. I'd finally talked to the man and he encouraged me to contact him." He sent Zadeh the proof at once. "And then I got the Big Call." Kosko was sitting in his office with four other people when the phone rang. "Zadeh said, 'If I can help you in any way, call me right up.' Well, this was just too much. It'd be like a little kid getting in touch with a movie star."

By 1985 he had moved on to a better-paying job and taken up Cheeseman's challenge. In early 1986, while preparing to teach a night course on fuzzy logic at the UC San Diego extension, he realized no one had developed a notion of fuzzy entropy. It was his breakthrough concept. He elaborated it, then chased down the issue of subsethood. These ideas appeared in his first two lectures.

Zadeh subsequently helped him set his dissertation in motion and served on his committee. Kosko earned his Ph.D. in 1987, USC quickly recruited him, and he began teaching there in 1988. He was now in position to devote full time to fuzziness.

THE 747 OF FUZZINESS

Is fuzzy logic the cocaine of science? Does it befuddle the scientific method? "That's an emotional point of view and I can appreciate it," he says. "It's the view of someone who finds out there's no Easter bunny."

For many people, he says, the imagined reality is simpler and more tractable than actual reality. So they cling to it.

However, when they "crack through this bivalence, like a Buddhist enlightenment thing, they find that fuzziness has a mathematical basis. In fact, there's a very rigid framework, a deterministic one." Kosko believes that Bayesians have spurned fuzzy logic because they have seen it as a cloud of generalities. But in fact "fuzziness doesn't fly like a balsa-wood airplane. It's a 747." This 747 is Kosko's own well-riveted structure, which he says recasts the entire field, vaporizes the major objections to it, and reveals its true relation to probability.

Kosko sought universal proofs, theorems which held for any fuzzy set, so he needed a framework which could unify them. He had to place them all in a single context. He decided to use a picture.

Pictures of math often enhance it. They can reveal larger patterns and deepen understanding. For instance, René Descartes invented analytic geometry to clarify equations, and circles show crisp sets reasonably well. How does one depict fuzzy sets? Circles do not work. Black and initially Zadeh had offered charts with concrete values like 6'2" on one axis and memberships on the other. They are vivid, but specific. A chart for height will not show temperature or speed. Kosko needed a larger structure. In 1971 Zadeh had suggested that a construct called a *hypercube* could model all fuzzy sets,[1] and Kosko developed this idea.

The hypercube smacks of analytic geometry. Descartes had placed a lattice over two perpendicular scales, rather like a Mercator projection of the world. Each point in the lattice had a unique position measured against the horizontal and vertical axes. The scheme worked just like longitude and latitude. For instance, if a point lay above 4 on the horizontal scale and across from 7 on the vertical scale, it was at (4,7). Every point had a two-number ID.

With a little juggling, such numbers can stand for memberships in a set. The result is the hypercube. Its scales extend only from 0 to 1, so it is a block rather than a boundless plane. When the set has two members, the hypercube is a two-dimensional square, and every possible set has its own point. For instance, the point at the lower left corner represents the set (0,0)—two zero memberships. This is the empty set. Kitty-corner at the upper right lies the point for (1,1)—two full memberships. At the upper left and lower right are (1,0) and (0,1), respectively. These four corners match and exhaust the true/false combinations in bivalence. In crisp logic, they are stars in a vacant sky. In fuzzy logic, however, the space fills in with a band of light. Points can represent the sets of (0.2, 0.8) or (0.15, 0.9) or (1/π, 1/√2)—anything. The square becomes a continuum. The crisp points bound it, and the fuzzy points flesh it out.

In two dimensions, the hypercube acts as a handy visual guide to fuzzy operations. For instance, it readily displays subsets. Imagine a set with memberships 0.2 and 0.7. As Figure 13 shows, it resides at the point (0.2, 0.7), called Set. Any point within the striped rectangle is a subset of Set, and any point outside it is not. For instance, (0.1, 0.6) is a subset, (0.9, 0.6) is not. The subsets stand out clearly and immediately.

Likewise, this construct depicts intersection and union in a straight-forward way. Intersection is simply the area where the two set rectangles overlap, as in Figure 14. Union, on the other hand, is the farthest extension of the sets, as in Figure 15.

The hypercube also shows the great felonies of classic set theory. If a set lies at (0.2, 0.7), its complement is at the fuzzy kitty-corner, at (0.8, 0.3). As Kosko observed, one can then take the set and its opposite and perform intersection and union on them. By the Laws of Contradiction

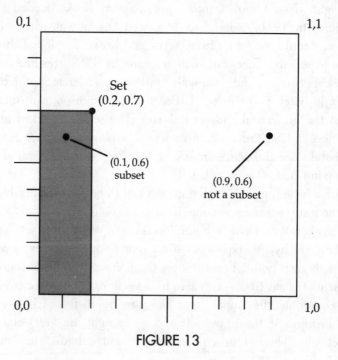

FIGURE 13

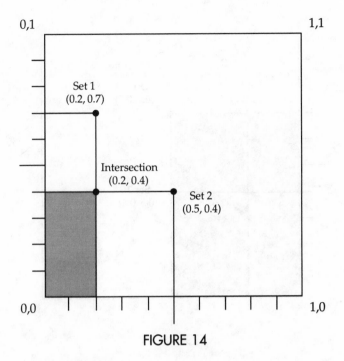

FIGURE 14

and the Excluded Middle, these operations must yield the empty set (0,0) and the full set (1,1), respectively. But here the intersection lies at (0.2, 0.3) and the union at (0.8, 0.7). Together, all four operations yield a rectangle centered on the midpoint of the hypercube.

Kosko says such crimes of classical logic are the essence of fuzziness. "Fuzziness is all about how much a thing and its opposite occur simultaneously. I call this 'boxing Aristotle into a corner.' The good news for Aristotle is that he is right somewhere: at the corners. The bad news is that he is right *only* at the corners. He is wrong infinitely many times."

What if the set has more than two members? Kosko added another dimension for each one. Every new member inhabits a dimension all to itself. For instance, a set of 0.6, 0.4, and 0.7 would need three dimensions and yield a point in a cube. A set of 0.3, 0.4, 0.1, 0.9, and 0.6 would require five and occupy a point in five-dimensional space. *Tall men* would have over a billion dimensions and be a point in a space no one can imagine. As a tool of visualization, the hypercube is a disaster with sets like *tall men*, especially compared to a chart. Yet it served Kosko's mathematical ends, for whatever a set's size, the hypercube let

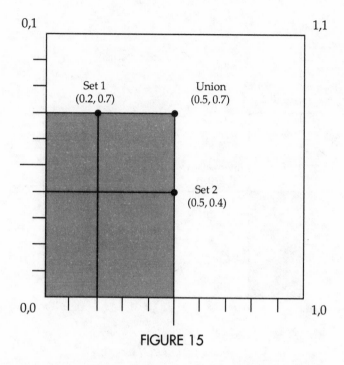

FIGURE 15

him conjure a space with the right number of dimensions and identify its point. Kosko calls this construct the *sets-as-points theory*. Each set has its point.

Kosko says fuzzy logic had seemed to be hanging out in the galactic emptiness, a shapeless nebula. The hypercube placed the same numerical gridwork on all fuzzy sets and let him develop mathematical proofs that applied to all of them. It also allowed him to rewrite some of the basic tenets of fuzziness.

EXPANDING THE REALM

"I personally think Zadeh made some mistakes," Kosko says. "One, he dealt with elementhood, in other words, how much an element belongs to a set." Like Max Black, Willett Kempton, and by then Zadeh himself, he saw the need to measure partial subsets, how much one set is part of another, how much *mugs* are *cups*. But Kosko went farther, and made partial subsets the base of a new structure. "The correct issue of fuzziness is not at the level of the element. It's at the level of the set," he says. "It's not how much the element belongs to the set. It's how much one set belongs to another set."

In October 1989, Kosko was still grappling with the issue. One day he went to the gym for a hard ride on the stationary bike, then returned home and meditated. For him meditation is no soft-brained Eastern twaddle. Rather, it is part of his rugged ethos: a challenge, an achievement. "You sit in the hot tub when the beta endorphins are high and just blank your mind as much as you can. Fifteen minutes will seem like an eternity the first few times. It's hard, but it builds self-control. I was a different person before I started doing it. It builds bulk, just like lifting weights. You can't cheat on this test. It's just you and your organism."

On this day, suddenly, he says, "with the high beta endorphins and in the midst of zazen meditation, pure blank mind, it just jumped into my head." He raced for his computer and began typing. A waterbed lay over in the corner. He imagined it aloft, suspended in hyperspace, and realized he could tie a string from himself to the bed and see where it touched. From there, he worked out a geometric proof: Probability is part of fuzziness.

The first step in this proof entailed extending fuzziness to subsets. Zadeh had explained fuzzy sets in terms of single elements, each with a degree of membership. However, single elements also qualify as sets, and here Kosko found his lever. For instance, as an element, Amazon is wholly in *long rivers*, as its membership of 1.0 attests. Therefore, as a set, Amazon is a 1.0 subset of *long rivers*. By the same token, as an element, Danube is in *long rivers* to 0.7 extent. Hence, as a set, it must be a 0.7 subset of *long rivers*. It is a fuzzy subset.

If so, Kosko asked, why not go one step farther? Why not assess the "membership" of multimember sets in each another? If Danube can be a fuzzy subset of *long rivers*, then *tropical rivers* and *polluted rivers* should be fuzzy subsets of *long rivers* too.

Determining subsethood—how much one set belongs to another—is very direct. It involves a simple percentage. One assesses what percent of one set lies in the other. For instance, suppose one set is the *swimmers* in a small apartment building and a second set the *golfers*:

	TINA	MIKE	GREG	SUE	TOTAL
Swimmers	0.8	0.7	0.9	0.2	2.6
Golfers	0	0.9	0.4	0.8	2.1
Both	0	0.7	0.4	0.2	1.3

There are 2.6 *swimmers*, 2.1 *golfers*, and 1.3 people who are both. Since 1.3 of the *swimmers* are *golfers*, *swimmers* is a 1.3/2.6, or 0.5, subset of *golfers*. Likewise, 1.3 of the 2.1 *golfers* are *swimmers*, so *golfers* is a 1.3/2.1, or 0.62, subset of *swimmers*.

This concept broadens the base on which fuzzy logic rests. It shows that the principles governing multimember sets control individual memberships as well. The borders of fuzziness expand.

And in this new and larger empire, fuzzy logic suddenly holds sway over many smaller kingdoms. "Probability gets sucked up by subsethood," Kosko says. "Entropy gets sucked up by subsethood. The classical theory of fuzzy sets gets sucked up by subsethood. It's a very powerful notion."

THE FUZZY MATRIX OF PROBABILITY

Kosko went on to show that one of the tributary domains was probability. Fuzzy logic forms its matrix, the framework within which it exists.

He demonstrated this fact by applying the fuzzy subsethood theorem to crisp subsets. For instance, flipping a nickel six times might yield the following sets:

Tosses	1	1	1	1	1	1	= 6
Heads	1	1		1			= 3
Intersection	1	1		1			= 3

Here, *heads* is a complete subset of *tosses*, since all heads are tosses. But Kosko then asked, "How much is *tosses* a subset of *heads*?" In set theory terms, it is like asking how much all Italians belong to all Venetians or how much South America is a part of Brazil. It sounds odd. But the subsethood idea provides the answer: *Tosses* is a 3/6, or 50 percent, subset of *heads*. As it happens, this result exactly matches the probability of *heads*. Kosko went on to derive Bayes's Theorem from subsethood.

Kosko's geometrical proof is, he says, mathematically firmer than this arithmetical description, and it rests on the hypercube. Where are the probabilities in the hypercube? Since they must add up to 1, in the square they form a diagonal line running from (0,1) to (1,0), as in

Figure 16. Similarly, in a cube, they form a slanting triangular plane. They are a sliver of the fuzzy sets possible.

Before the fuzzy subsethood theorem, Kosko says, probability rested on relative frequency as an axiom, beneath which lay only intuition. His geometric proof unveiled a deeper structure based on fuzziness. Now, he says, the probabilist's "axioms are the fuzzy theorist's theorems. What shakes them up is that what has been taken previously as naked definition now has got back as theorem," that is, as proof.

He is referring to the modern rule of priority in math: If one field's assumptions are another's proofs, the first is a part of the second. Indeed, discovering such foundation structures has been a hallmark of contemporary mathematics. Examples abound. Multivalued logic and fuzzy logic contain classical logic. Non-Euclidean geometry includes classic geometry, and fractal geometry embraces both. In physics,

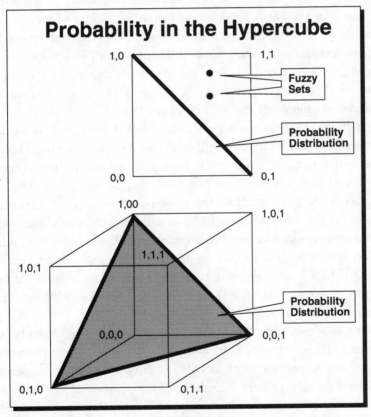

FIGURE 16

unified field theorists have generalized the four basic forces of the universe to three, and continue the quest for just one. Indeed, in fuzzy logic itself, subsethood includes elementhood.

Since fuzziness contains probability as a special case, Kosko says, everything probability does, fuzziness does as well. The Bayesian claim to superiority falls apart. It can hardly exceed fuzziness if it is fuzziness.

"The identification of relative frequency with probability is cultural, not logical," Kosko concludes. "This may take getting used to after hundreds of years of casting gambling intuitions as matters of probability and a century of building probability into the description of the universe. It is ironic that to date every assumption of probability—at least in the relative frequency sense of science, engineering, gambling, and daily life—has actually been an invocation of fuzziness."

Other theories, such as Michio Sugeno's fuzzy measures and Dempster–Shafer belief functions, subsume probability as well.[2] But George Klir says Kosko's proof is "amazing, because it comes so naturally." He cautions that we still do not fully understand the overlap of fuzziness and probability, but adds that Kosko's proof poses a major challenge for Bayesians. If they "claim probability can do what fuzzy logic can," he says, "then they should show that fuzzy sets can be captured in probability theory. I don't see that."

Yet Judea Pearl says such proofs suggest that fuzzy logic is too lax. "I'm worried that it doesn't instill sufficient discipline in what a person is permitted to do, that it is essentially equivalent to a vacuous theory," he says. It is easy to create a construct that derives probability as a special case, he notes. Just take the axioms of probability and relax them arbitrarily. You now have more freedom and you feel great. The problem is that counter-intuitive conclusions may arise. With fuzziness, he adds, "Everything is permissible. If I say probability, they say, 'Special case of fuzzy.' If I say belief functions, they say, 'Special case of fuzzy.' If I do something sloppy, they say, 'All right, special case of fuzzy.' What is not a special case of fuzzy?"

Klir's challenge, however, remains unanswered. If probability can surpass fuzziness at every task, Bayesians should be able to prove that fuzziness is a special case of probability. No such proof has appeared.

Even if Bayesians accept the validity of the proof—and Pearl does not deny it—it may or may not impress them. For the Bayesians are divided about the value of deductive structure to the field. On the one hand, many praise the axioms of Leonard Savage, and Dennis Lindley feels that some system of axioms is desirable and that "the lack of them must count against the alternatives to probability."[3] One might expect this camp to retreat somewhat before Kosko's work. On the other hand, E. T. Jaynes feels Bayesianism has a feeble axiomatic base and by 1986 had discounted the whole axiomatic method. Objectivists and mathematicians, he said, start from the ground up. Bayesians leap into the center and work up and down, like problem-solvers in real life. "Mathematicians sometimes dismiss our arguments as nonrigorous," he says, "but the reasoning of a physicist, engineer, biochemist, geologist, economist—or Sherlock Holmes—cannot be logical deduction because the necessary information is lacking."[4]

But Kosko did not stop with subsuming probability. Fuzzy logic, he says, embraces even wider fields, including arithmetic.

FROM ENTROPY TO ARITHMETIC

He began with one of the prize accomplishments of Bayesian probability, Claude Shannon's information theory, and proceeded to make its basic concept over in fuzzy terms.

Entropy has become famous as a Big Word, a term that falls nonchalantly from the poseur's lips, yet defies comprehension. The German physicist Rudolf Clausius coined it in 1865 in connection with the Second Law of Thermodynamics. Its opacity thus stems partly from its artificiality. Like a euphemism, it lacks human resonance. In fact, it simply means disorganization.

Where does disorganization lurk? A bedroom or a desk drawer can be disorganized. So can a termpaper or a speech. And in fact these examples reflect the two main arenas of entropy: physical and informational.

The physical side has received more attention, partly because of the eerie upshot of the Second Law of Thermodynamics. It holds, briefly, that in a closed system, like a jar, disorganization can only increase. Heat the bottom of a jar—that is, open the system—and the lower levels of air warm up and it starts to circulate. It is more organized. But remove the jar from the stove—close the system—and eventually a uniform temperature prevails again. The air molecules become less organized.

Every part of space resembles every other part. Eventually, according to this law, the entire universe will become homogeneous in a similar way and lapse into "heat-death." Thus, the Second Law says: Things disintegrate.

Information is a kind of physical organization. It is patterns of ink in publications, fired pixels on a TV screen, amino acids in DNA, ons and offs in digital computers, neural links in the brain. When such patterns seem disorganized, said Claude Shannon, in founding information theory in 1948, we have uncertainty. Thus, entropy is also uncertainty.

In fact, Shannon defined information very narrowly, not in terms of spoken words, but of bits flowing through a digital computer. A *bit* is the smallest unit in a computer and can be only on or off. It is bivalent, so Shannon could use probability to describe it. He also relied on the Bayesian notion of expectation. Imagine a series of bits arriving through a data channel. If every bit so far has been an *on*, the observer naturally expects the next one to be an *on* also. There is organization, a pattern. If the incoming bit actually is an *on*, Shannon said, it imparts little information. The entropy or uncertainty was low. The observer has already guessed it.

Suppose, however, that the stream of bits is random. In that case, one has no idea what to expect, and uncertainty is high. So if the next bit is an *on*, it conveys much more information. It has resolved more uncertainty. Thus, Shannon said, in a famous definition, that information is a message that resolves uncertainty.

Kosko addressed fuzzy entropy with a simple question: How fuzzy is a fuzzy set? He observed that fuzziness is a kind of uncertainty and hence a kind of entropy. A crisp set, one at the corners, is certain. Its entropy is zero. For instance, the set of odd integers between 0 and 10 has five members, all with 1.0 membership. There is no doubt about their status.

However, a fuzzy set can contain points in the midst of the hypercube and their membership is less certain. The more easily an item rounds off, the more certain we are of it, one way or another. For instance, the set of *small* integers between 0 and 10 might have members with values of 0.9, 0.5, and 0.2. Which is least certain? The 0.9 very easily rounds off to *small*, and the 0.2 to *not-small*. But the second membership, 0.5, won't round off either way. Is it *small*? *Not-small*? We have no idea. Likewise, a half-filled glass has a 0.5 membership in the set of full glasses, for we are utterly uncertain about whether it does or doesn't belong. Here, at dead center of the cube, there is maximum uncertainty. We are at paradox point, the realm of conundrums like "I am lying."

From this base, Kosko developed his fuzzy entropy theorem, which is essentially the violations of the Law of Contradiction divided by the violations of the Law of the Excluded Middle.[5] This scofflaw theorem, he says, also yields a basic derivation of Zadeh's rules for determining complement, intersection, and union. Kosko took it even further, to the point where he could claim that all addition is fuzzy entropy, and so is all arithmetic. "You can eliminate addition in favor of fuzzy entropy operations," he says.

THE PROBABILITY NEXUS

Kosko's vision of fuzzy logic, however, does not simply dwarf all other schemes into triviality. In June 1992, he announced another proof, one that dealt with fuzzy systems like Mamdani's steam plant and that surprised almost everyone.[6]

Zadeh had noted in 1971 that fuzzy systems amount to mappings between hypercubes. A system resembles a cluster of cubes in space, with arrows linking sets in one cube with sets in another. For instance, in the Mamdani steam plant, one cube could be an input family of fuzzy sets like pressure error, and arrows could lead to an output cube like heat change. These matchups are functions—means of changing one piece of information into another, such as formulas or IF–THEN rules.

Kosko's proof said simply that the math of probability could also describe these systems. It could render fuzzy membership functions— the triangles, bells, and trapezoids—as conditional probability densities. And it could carry out defuzzification, or locating the center of gravity of the final geometric shape, by finding a conditional mean. "There's not one step that's outside anything done with probability," he says.

Yet he also announced caveats. Though probability can describe, say, positive small, it is more complicated. "Fuzzy is mathematically simpler," he says. So if his new proof is valid—and some like Zadeh urge caution—it shows that the Bayesians had some basis for their intuition about their claim, since the math of probability handles fuzzy systems better than most fuzzy theorists realized. On the other hand, it seems to undermine any Bayesian claim to equal or superior performance.

Moreover, the proof affects only the underlying math. The Bayesians deploy a different method as well. They demand math models, over-arching numerical descriptions which can cost time and money. Fuzzy systems do not need this apparatus since fuzzy rules provide a shortcut around it, yet Kosko says these rules are "beneath the dignity of a statistician." Fuzzy systems simplify the method as well as the math. "I think fuzzy systems are the most convenient way known to approximate nonlinear systems," he says. "They are simply new ways to compute conditional means. The number 4 has many ways of being computed."

Kosko emphasizes that the proof applies only to fuzzy systems, between-cube mappings, and not to fuzzy sets. "Tom is tall" remains fuzzy. He also notes that probability cannot describe every fuzzy application. "There are very elaborate systems that use more than these simple sums of rules," he says, such as fuzzy error-correcting codes, that probability will not render at all.

The probabilists, Kosko says, were simply too ignorant of fuzzy systems to realize the irony of their position. "Peter Cheeseman and the rest of them fall down because they try to argue against the applications, without knowing what they are."

Mathematically, methodologically, and conceptually fuzzy systems are plainly distinct from probability systems. And their direct link to human words and thoughts gives them a power probability must contort itself to mimic. "There's a formal reason why it's easier to do these systems as fuzzy: consistency," Kosko says. "We model the whole world as fuzzy, so why change when you're building a fuzzy system?"

There is no better illustration of this idea than fuzzy expert systems, the fuzzy answer to the AI quagmire.

11

FUZZY DELPHI

■

Something like $100 billion worldwide has gone into AI and you can't point to a single AI product in the office, home, automobile, anywhere. With neural networks, it's approaching $500 million, and you can't point to any products and you probably won't be able to for some time. Fuzzy techniques have gotten almost zero dollars from the government, and you can point to a lot of products, and will be pointing to a whole lot more.

—BART KOSKO

AI has not been able to attain the objectives it set for itself. And I place the blame on its commitment to classical logic.

—LOTFI ZADEH

In a matter of two to five years, you'll find most expert systems will use fuzzy logic.

—LOTFI ZADEH, JUNE 1991

■

Modern computer history has a single Great Disappointment. It is AI, the attempt to bottle feats of mind in a machine. It began with much fanfare and apparent success in the 1950s, spurring glamorous visions of machines that could think. Government and universities all over the world pumped vast sums of money into it, and computer scientists built shimmering reputations on it. And by the 1980s, it had all stalled out.

Why?

Part of the reason lay in a fact Friedrich Nietzsche pointed out over 100 years ago: "Thinking which has become conscious is only the smallest part of it, let us say the most superficial part."[1] Indeed, the mind is at once so familiar we feel we understand it, yet so mysterious it leaves us ultimately groping. We know it like the normal driver knows a car. He understands the pedals and buttons and levers, but almost nothing about what goes on under the hood.

AI investigators assumed nothing went on under the hood. They saw conscious thought as all thought. Hence, they believed chess would be difficult to program and vision easy, because chess demands intense concentration and vision none at all. The scale of this delusion is hard to convey. Chess programs can now challenge world champions, but the best computer retinas remain almost blind. Vision requires massive information processing. We are just not aware of it.

Moreover, though the human brain is the most complex object on earth, AI workers insisted on imitating it with one or two techniques—and with symbol-based reasoning in particular. They were like cartographers trying to map London by hearing alone. As MIT's Marvin Minsky, a doyen of the field, says, "The big problem in AI is people joining one party and rejecting almost everything else." It now seems impossible that anyone will reduce intelligence to a single model, and of all modern scientists, AI researchers could most profit from the wisdom of the blind men and the elephant.

They plainly spurned fuzzy logic. For instance, in 1980, Lotfi Zadeh submitted a paper to the First Annual National Conference on Artificial Intelligence. Reviewers accepted 63 out of 200 submissions, but returned his. The paper was not inferior. Rather, explained one reviewer, his work was not well-known in the AI community, so there might be antagonism to it. He added that most current work on natural language processing did not involve interpreting imprecise words.[2]

Ten years later, Zadeh said, "Now the community that could benefit most from fuzzy logic is the AI community. And yet the AI community is the one that is the most hostile. The question is: Why? Because the AI community is committed to classical logic. And once this kind of commitment is made, it is very difficult to put aside."

This attitude may be slowly changing. Nils Nilsson, a Stanford professor who in 1989 called fuzzy logic a "temporary idiosyncrasy,"[3] recently told Zadeh he was going to employ it in a project. And if the pall of ignorant scorn is lifting, at least part of the reason is fuzzy expert systems.

DELPHI IN A BOX

Expert systems are canned authority. The first AI program—and the primeval expert system—proved theorems in symbolic logic. Logic Theorist, invented in 1956 by Allan Newell and Herbert Simon, began with premises and used a few rules of inference to combine them or alter them. Such rules would yield a proof if applied in the correct order. The key lay in finding that sequence.

Newell and Simon could not just proceed randomly, testing out each result, because it would take far too much time. Instead, they used *heuristics*, a somewhat forbidding term that means rules of thumb, hunches. Newell and Simon gave the progam simple guidelines: Try this approach, and if it doesn't work, try this other one, and so on.

Heuristics are very different from algorithms. Algorithms specify each step cleanly. Heuristics offer guesses, and their outcome is uncertain, unpredictable. Algorithms are tidy, heuristics are not. But they excel at coping with the complex and unexpected. Good heuristics smack of intuition, and of course people use heuristics to handle problems all the time. It also worked for Logic Theorist. On August 9, 1956, the program yielded its first complete proof, of a theorem in Russell and Whitehead's *Principia*. It would later prove another *Principia* theorem more briefly and elegantly than the authors had.

The next year, Newell and Simon unveiled General Problem Solver, which applied broader means of reasoning to solve problems beyond the sphere of logic, such as those involving games and puzzles.

These programs sparked much excitement. Computers could think! But this general problem-solving ability turned out to be surprisingly useless. People reason with knowledge. It is another self-evident fact that has been amazingly unself-evident. Heuristics depend on the problem at hand, and the more we know about it, the better our hunches and the faster we will reach them. A top-flight chess program may consider 5 million moves in the time a grand master considers three, yet the grand master can win because she senses the most promising avenues at once. Indeed, Simon and others discovered that chess masters do not reason any faster than novices, but rather have learned so many board positions and strategies that they grasp the heuristics better.[4] By relying on sheer reasoning prowess, Newell and Simon forced their program to start from scratch with every problem.

It is a revelation with major implications for theories of human intelligence: Knowledge may be inseparable from intelligence. The more you know, the smarter you become.

It also raises serious questions about logic itself. If the mass of baseline knowledge called common sense is so important for solving problems, what are we to make of the pristine operations of logic? For instance, consider:

All oak trees have acorns.
This tree has acorns.
This tree is an oak tree.

People are much more likely to agree with this fallacious logic than with

All pro basketball players are very tall.
Bob is very tall.
Bob is a pro basketball player.

The two are logically identical, but our knowledge about the background differs. Only oak trees have acorns. But we know many very tall men are not pro basketball players, so counterexamples quickly spring to mind. Knowledge dominates everyday reasoning.

Computer scientists soon began to put this precept into practice, boosting AI with knowledge. One of the first was Edward Feigenbaum of Stanford, who coined the term *knowledge engineering* to describe this goal. In the 1960s, he, Joshua Lederberg, and Bruce Buchanan developed DENDRAL, the first expert system. A device called a *mass sprectrometer* would break organic molecules apart and weigh the pieces. DENDRAL contained rules for divining the molecules' structure from these weights. Scientists fed in the weights and it suggested possible shapes. The knowledge was the expert part of the system, and it consisted of rules from scientists.

However, the pacesetter was MYCIN, begun by Edward Shortliffe in the mid-1970s to diagnose illnesses. It became the model for future expert systems, and it differed crucially from Logic Theorist. It had three main parts—knowledge base, database, and inference engine—of which only the latter descended from Logic Theorist (see Figure 17).

The knowledge base and database resemble the first two premises of a syllogism. The knowledge base contains general principles like "All men are mortal," the expertise doctors possess. Shortliffe originally placed some 500 IF–THEN rules here, each corresponding to a piece of medical knowledge. For instance:

IF
1. The stain of the organism is gram-negative, and
2. The aerobicity of the organism is anaerobic, and
3. The shape of the organism is a rod,
THEN
There is suggestive evidence that the organism is *Bacteriodes*.

The second part, the database, contains specific information about the patient, like "Socrates is a man." It includes age, sex, medical history, and various test results.

The third part, the inference engine, applies the database facts to the rules in the knowledge base and tries to reach an answer. If it needs more facts, it can ask the doctor at the computer screen, yet if the doctor cannot supply this information, MYCIN will continue, doing its best.

As with Mamdani's steam plant, the power in MYCIN lies in the knowledge base, the storehouse of rules. Shortliffe wisely segregated it completely from the inference engine, so he could augment or alter it without overhauling the entire system.

Expert System

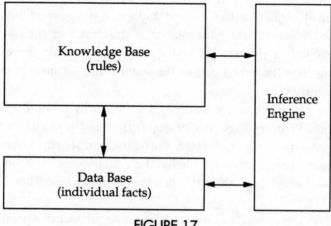

FIGURE 17

MYCIN had focused on blood and meningitis infections. But the MYCIN workers found they could detach its inference engine from the knowledge base and use it in other fields. This floating inference engine was called an *expert shell,* and the first one was EMYCIN, for Essential MYCIN or Empty MYCIN. It led to programs like PUFF, for pulmonary problems, and HEADMED, for psychiatric disorders.

MYCIN took over 20 man-years to make, but it is a success. In one formal investigation, a team rated MYCIN's drug prescriptions accurate 65 percent of the time, while human physicians were accurate 42.5 to 62.5 percent of the time.[5]

Since then, other expert systems include PROSPECTOR, a Bayesian program that advises geologists searching for ore (and in 1980 located a molybdenum deposit in Washington state), and R1, which devises possible configurations for Digital Equipment Corporation's VAX computers and has saved the company perhaps $20 million per year.[6]

UNCERTAINTY IN EXPERT SYSTEMS

When logic ventured into the real world with expert systems, it had to face problems it had long brushed aside. For instance, medicine is not a crisp field. Diseases have fuzzy definitions. For instance, osteoporosis is "an absolute decrease in the amount of bone to a level below that required for mechanical support,"[7] and it comes on by degrees. Pain is a basic symptom, yet patients cannot quantify its type or degree. They express it in fuzzy terms. ("It's a dull pain that hurts a lot.") Moreover, it can be very difficult to diagnose a disease exactly, and even when doctors know the ailment, they cannot always discern the precise treatment. Hence, the makers of MYCIN gave its variables and rules numerical *certainty factors*, or CFs. Depending on how they combined, the CFs yielded an estimate of the reliability of the diagnosis.

CFs, however, have limitations. They don't fire a rule partially, but instead generally tell whether to fire it at all. They are clearly not the only way to express uncertainty. For instance, PROSPECTOR employs Bayesian probability, reasoning from evidence of deposits to hypotheses that deposits exist. Some expert systems have used Dempster–Shafer theory, and some use no means of registering uncertainty at all.

The debate about the effectiveness of these methods has been long, loud, and acrid. The Bayesians dismiss all other approaches. But Glenn Shafer says that expert systems do not handle Bayesian reasoning adeptly "and many AI theorists would conclude from this that probabilistic reasoning has little role in genuine intelligence."[8] Ian Graham, a British consultant, observes that "people are very bad at estimating probabilities, which makes the Bayesian approach inappropriate for systems which will be used by non-statisticians."[9] Graham adds that the best approach hinges on the type of uncertainty in question, and that future expert systems should use several methods, since the temptation is to trot out the same one for all uncertainty.

FUZZY EXPERT SYSTEMS

Fuzzy expert systems work very much like Mamdani's steam plant, firing rules partially and fusing them for a result. Hans Zimmermann invented the first in 1977, a program that helped banks evaluate loan applicants and analyze stock portfolios. "We put about ten man-years into empirical research in fuzzy sets," he says. "And the last step of this project was a big field study concerning the evaluation of credit business around Aachen, and after this was finished, people said, 'Why don't you put it in an expert system?' " They did, and the program, called ASK, evolved into a commercial product which is now on the market.

Another early fuzzy expert system was CADIAG-2, created by Peter Adlassnig of the University of Vienna in the early 1980s to perform medical diagnosis. Based on massive patient histories at the Vienna General Hospital, CADIAG-2 used fuzzy sets to represent the intensity of symptoms, the degree to which symptoms indicated disease, and the degree of belief that patients had particular diseases. In 500 test cases, the program reportedly scored a success rate of 93 percent.

Fuzzy logic addresses several problems with current expert systems. One, of course, is representing fuzziness itself. In 1983 Zadeh noted that since most knowledge in an expert system is fuzzy, most of its facts and rules are fuzzy. For instance, a rule-based system for playing poker might have the following rule:

> If your hand is *excellent* then bet *low* if the opponent *tends* to be a *conservative* player and has just bet *low*. Bet *high* if the opponent is not *conservative*, is not *easily bluffed*, and has just made a *sizable* bet. Call if the pot is *extremely large*, and the opponent has just made a *sizable* bet.[10]

Existing systems, he noted, ignore the fuzziness of such information. For instance, MYCIN might conclude: "Rodney has a duodenal ulcer (CF 0.3)." But the meaning of the CF is unclear. Does Rodney have a 0.3 chance of having a duodenal ulcer, or does he have one to 0.3 extent? Fuzzy theory, Zadeh said, can clear up such problems.

Though representing fuzziness is important, fuzzy expert systems also employ the Mamdani structure, which is arguably even more important. Most conventional expert systems are decision trees, or gardens of forking paths. They make one yes/no decision, then move on to another, and another. In general, according to Kosko, a decision tree has these drawbacks:

1. It is prone to closed loops, in which the software reads both IF A THEN DO B and IF B THEN DO A and spins off to infinity.
2. It is slow, and the more rules it has, the more koalalike it becomes. In the mid-1980s, for instance, some estimated that the power available to computers set a practical limit of a few thousand rules per system. According to Thomas Knight of MIT, that meant no

program could ever guide a robot across the room and have it set the table for dinner.

3. It does not use all the data. The paths are independent. Instead, we want them to blend, but without causing closed loops.

4. It is not combinable. "A tree plus a tree doesn't equal a tree," says Kosko. "This is the curse of AI. If you add three, four, or many expert systems, you will likely have a structure peppered with closed loops."

Fuzzy expert systems merge the rules. "You don't simply evaluate the rule and take one path or the other," says programmer Marlin Eller of Microsoft. "You sort of take both, but you take more of one than the other based on what that rule is. You sort of run all the paths of the tree simultaneously and do a weighted average of the results." Instead of forking paths, the system takes many paths at once. It works in parallel.

A MYCIN FOR MACHINES

The benefits of this approach become clear as one observes fuzzy expert systems in action. Tadashi Katsuno is an assistant manager at Omron Corporation, which developed its fuzzy medical diagnosis expert system between 1984 and 1986. A doctor elicited symptoms from a patient and fed them into the software, which could handle about 300 symptoms and 300 diseases. It analyzed the symptoms and concluded that the patient likely had, say, diabetes.

Katsuno found that one bonanza lay in streamlining the rules. The proliferation of rules has reached gridlock for expert systems, and fuzzy logic slashes them. "The number of rules required becomes 100 or 1,000 times smaller," he notes. (Michio Sugeno says it cuts the rules by at least 90 percent. The precise amount is unclear, but it is huge.)

Fuzzy expert systems yield other boons. They telescope development time, automatically combine the knowledge of many experts, respond where conventional expert systems stand silent, and are easy to update.

For instance, Omron developed an expert system to diagnose problems in a large device called a gear shaver, which cuts gears as big as 2.5 meters across. It was a MYCIN for this machine. If the gear shaver showed a symptom such as sudden difficulty in rotating the operating handle, one could feed this information into the expert system and it would reply with a diagnosis, such as breakdown in the brake, fan, or sharpener.

To build this software, Omron went directly to 10 maintenance people. Over a month and a half, these authorities set down their knowledge of the extent to which certain defects in the gear shaver caused symptoms.

The conventional expert system can accept the opinions of many people, but when they differ, as they typically do, they "must be put into one standard opinion, one solution. You must study the data and make it into a certain structure," Katsuno says. "Otherwise the system cannot tell what to do. And this is usually a job that takes months. This cannot be done by the operators at the factory."

Here, however, the operators simply plugged their information into the system, which automatically blended the rules from all of them. It took just half a day. "And that is very, very different compared to the conventional." Katsuno says it took one and half months to start running this system. "That is at least four times shorter than other expert systems, which would take about six months." Moreover, he says, "The maintenance time for that machine was reduced to 24 percent, by using this system."

In addition, the fuzzy expert system speaks where traditional ones are mute. For example, he says, if a conventional system needs ten pieces of information to diagnose a broken sharpener, but only receives six, it will not respond. But the fuzzy system might say the probability of a broken sharpener is 35 percent. "So this is another point," Katsuno says. "Every piece of information is not necessary. The system always gives you some answer."

However, he adds with a laugh, "At the very beginning the answers were not very good. Sometimes it gave more than one answer. For example, number one 50 percent, number two 20 percent, number three 1 percent. That happens. But anyway, if you find something strange or not very good, you add new data to the system and the system takes care of rebuilding the knowledge base automatically. So the knowledge engineer is not needed."

In other words, the system is easy to update, an important factor. "Once the system starts operating," Katsuno says, "the products to be manufactured may change and the aging of the machine will produce different types of failure, so people may discover more and more new problems." With conventional expert systems, the firm must call in the knowledge engineer, explain the new snafus, and ask her to modify the system. But in the fuzzy expert system, he says, "You just add new information to the system and tell the system to recombine or regenerate the database. That's all. It's really very, very easy to update the system while running."

Bob Lea, an aerospace engineer who has helped create fuzzy expert systems at NASA, has noticed the same bonuses. "I think with the work the Japanese have done, people are saying we ought to think about fuzzy logic a little more and they're not just grinning and chuckling about it," he says. When they do start examining it, he says, "I think they'll find it's one particular tool that'll make expert systems work. Quite frankly, they haven't worked very well so far."

Many problems have beset them. "The rule sets are too sparse and don't cover everything," he says, "or the problem gets blown up to the point it's unmanageable. With fuzzy logic, you can state rules in expert systems the way people think about them." He too notes they use fewer rules and can resolve contradictions in the rules experts provide.

By now, numerous fuzzy expert systems have appeared all over the world. Though many are experimental, they include PROTIS (treatment of diabetes), EXPERT (rheumatology), SPHINX (medical diagnosis), SYNTEX (hospital management), and SPERIL I (assessment of earthquake damage to buildings).

Hitachi sells an all-purpose fuzzy expert shell, called ES/KERNEL. "It's the best-selling expert system development system," says Akira Maeda, a researcher at the Systems Development Lab in Kawasaki. "There are several thousand sold." It is a powerful tool, one of the major commercial achievements in this area, and it is seeding an array of fuzzy expert systems in Japan. Among them, says Hitachi, are expert systems that

- Advise on creating the ideal façade for shops.
- Help determine whether to open a store.
- Help formulate schedules for flight crews.
- Diagnose problems in automobiles and copying machines.
- Help create advertising strategy proposals.
- Help make engineering decisions on girders in bridges and buildings.
- Help choose the best construction methods.

All these programs aim to save money, but some of the most intriguing seek to harvest cash directly.

MONEY MACHINES

"An expert system is nothing but a fuzzy system," says Michio Sugeno, who worked two years with Yamaichi Securities developing its fuzzy expert system for computer stock trading. The challenge of such an enterprise is perhaps exceeded only by its potential payoff. The stock market is a complex system that reacts to an array of other complex systems, and a program that can foretell its behavior is an outright money machine, an ATM in the corporate backyard spewing out endless cash.

Sugeno formed a committee of 30 people inside Yamaichi, who interviewed more than 10 experts and collected some 600 rules. They then refined the rules through discussions with other individuals and developed three kinds of database, relating to (1) the stock market, (2) companies, and (3) sources of money, such as banks. Once they had the software in place, the group tested it against past data from January 1987 through January 1989. Black Monday occurred during this period, and one week before it, the system gave the red alert and sold all the stock. "So it did not get damaged by Black Monday," Sugeno says.

Despite this success, the software later faltered as the Nikkei Stock Average began its drastic slide, losing over half its value from the end of 1989 through spring of 1992. Only a program sensitive to the larger society could have predicted this collapse, and it shows that money does not yet grow on trees.

MYCIN took a long time to develop, but at least it addresses a fairly stable field. The rules of medicine remain pretty much the same from year to year. But what about an area where the rules are changing constantly? Can expert systems handle rapid flux?

Hitachi has developed such a program for short-term bond trading. It is not an expert system in the sense of handing down opinions from on high, dispensing oracular wisdom. It is instead an advisor: a fuzzy decision-support system.

Its author is Chizuko Yasunobu, the wife of Seiji Yasunobu, the codeveloper of the Sendai subway. A delicate woman with a soft voice and intelligent manner, she explains her program on the screen in quiet, complex English sentences which articulate like those in a book.

Whereas the Yamaichi software identifies long-term trends, she says, her program spots short-term trends. Fuji Bank has now employed it for one year—not in simulation, but in daily use. (Unlike U.S. banks, Japanese ones can trade in stocks.) She estimates it has earned the bank ¥100,000,000 per month (at ¥130 per dollar, about $770,000), though a Japanese financial newspaper has placed the figure several times higher.

This software took two people six months to write. Yasunobu herself spent two months in a bank learning the task, then devised 150 rules. After reading John J. Murphy's *Technical Analysis of the Futures Market*, she added 50 more.

The program had to surmount several obstacles. Good rules change rapidly in this arena, and even experts lack the knowledge to navigate without setback, so they must constantly update their information. Trial and error inheres in the task. As a result, the program lets users define the fuzzy rules for recognizing patterns. For instance, a key pattern is a bull trend, a rise in prices. Buy early in a bull trend and one can ride it escalator-like to the top, selling out at a profit. This decision-support system analyzes such factors as the moving average— the average closing price over three days—and can highlight a bull trend before the trader becomes aware of it.

Because the rules are always changing, the program also researches new rules. It lets traders test their own rules against historical data, a popular feature, and evaluates rule performance. For instance, the bull trend rule has 60 percent accuracy. "It tells you the chance that the rule is correct. This is important," she says, with understatement.

Although traders are currently using the system only for bond futures, one could easily extend it to stocks and foreign currency exchange. "We are trying to widen the range," she notes.

FUZZY DATABASES

In 1978, Brian Gaines observed that some of the first computers found applications in exact areas such as accounting. Databases arose from this background and, he notes, most of them still require precision.

So they are brittle. For instance, if a firm wished to find which of its young employees showed promise, it might issue a conventional request like: "List salespeople under 25 years old who sold more than $40,000 worth of goods last year." This search would miss the 26-year-old who sold $100,000, as well as the 19-year-old who sold $39,000. It would create crisp sets and ignore noteworthy fuzzy information at the edges. A vaguer request would be handier and more useful: "List the young salespeople with a good selling record."

Maria Zemankova, who has written a text on fuzzy databases, says, "It's not difficult to build these systems." She notes that a fuzzy database exploits vagueness in three different ways.

First, it lets data be fuzzy. Instead of storing age as 30 years old, it might express it as *old* with a fuzzy membership of 0.6. Instead of stating a place of residence as "Bethesda," it could call it "Washington, 0.9; Baltimore, 0.6; and Annapolis, 0.4." Zemankova says this approach "allows you to deal with the uncertainty of the values as well as vagueness of linguistic terms such as *old* or *rich*."

Second, it permits fuzzy retrieval. When searching for 30-year-olds with green eyes, the fuzzy database can also identify 28-year-olds with green eyes and note the lower degree of match. "This is very important when looking for applicants for a particular position or discovering some phenomenon in a database," she says. "You don't want a strict query, because when you don't satisfy it, you go from complete satisfaction to zero satisfaction. Fuzzy allows less and less satisfaction. If there are no precise matches, it lets you know, 'This is not exactly what you're looking for, but it's the best I can do.'"

Third, it allows a fuzzy query language. "You can build, on top of a database, a fuzzy rule system that enables you to do nontrivial query answering," she says. "So you can have a fuzzy deductive database or a fuzzy expert system which softens the brittleness problem."

Thus far, Zemankova notes, fuzzy databases have rocked quietly back and forth in the doldrums. Yet the Japanese have translated her book and "I'm worried about this, because the United States is the world leader in information technology. And if the Japanese start producing fuzzy information systems, we'll lose the leadership in one of the last strongholds of computer science we still have."

THE SHAPE OF A CAT

What is the shape of a cat? asked philosopher J. L. Austin (1911–1960). Does it change when the animal moves? Is it a broad outline or as fine as the serried edges of each hair? Whatever the philosophical answer, human beings need only the general contours. For us, a cat is a pattern and like most other patterns, it is fuzzy.

Pattern recognition is another task the brain carries out swiftly, ceaselessly, compulsively, beyond our awareness. Show people alphabetic characters arranged randomly on a page and they will find patterns in them. Hang four black three-quarter circles correctly and they will see a white square that doesn't exist. Show them a stack of lines broken just right and they will read the trademark letters IBM. We find patterns because the brain forces order on experience. An organized world is easier to handle.

Pattern recognition comes much closer to normal human reasoning than logic does. We are gawky at rules, but outstanding at pattern recognition. It is essential for survival. We sense patterns of behavior and their meaning in ways that brazenly defy articulation. We use the grammar of our native language perfectly, yet can scarcely express its rules and must learn them in class. A good poker player can sense a winning hand without quite knowing why. Indeed, pattern recognition is almost undefinable, and in its broadest sense includes classification, insight, and almost every cognitive activity the brain carries out.

Because pattern recognition is fuzzy, it has long taxed digital computers. Yet even dull animals perform it effortlessly. For instance, pigeons are every bit as witless as folklore has it. Yet Stephen Lea, a professor of psychology in Exeter University in Britain, has shown that they can "fairly easily" learn to recognize a letter like A—whether handwritten or presented in a variety of typefaces. In a computer, this trick takes much power.

In 1973, Zadeh described fuzzy definitional algorithms that let computers outperform pigeons.[11] Rosch had shown that people perceive categories in a fuzzy way, and fuzzy definitional algorithms put this insight to work.

For instance, suppose one wants a computer to recognize ovals. What is an oval? What separates it from, say, a horseshoe or a circle? In the classic rationalist tradition, Zadeh took the parts of an oval asunder. He observed that

1. It always has a pool of interior space. It is closed, like a *D* rather than a *U* or a *C*.
2. It is always free from crossing lines. It is non–self-intersecting, like an *O* rather than an 8 or an *X*.
3. A straight line from any point inside the space to any other point will not stray over the edge. The shape is convex, like an *O* rather than a *B*.
4. It is symmetrical along two lines or axes, which are roughly perpendicular to each other. It is like an *O* or an *X*, rather than a *V* (symmetrical on one axis) or an *R* (asymmetrical).
5. Finally, one of these axes is quite a bit longer than the other. It is like an *O* or *X*, rather than an *o* or *x*.

If this analysis seems agonizingly self-evident, it simply proves the point. People are so good at pattern recognition that they don't need to think about it. An oval is one of Rosch's basic categories, and we recognize it without realizing how or why, but computers must build up from parts.

From this five-step definition, Zadeh devised the recipe for recognizing an oval. After each instruction below, the process moves on to the next unless it reaches "stop":

1. IF shape is not closed THEN shape is not *oval*; stop.
2. IF shape is self-intersecting THEN shape is not *oval*; stop.
3. IF shape is not convex THEN shape is not *oval*; stop.
4. IF shape does not have two *more or less* perpendicular axes, THEN shape is not *oval*; stop.
5. IF the major axis is not *much* longer than the minor axis, THEN the shape is not *oval*; stop.

These steps have interesting features. For instance, the first three are crisp. Zadeh could have made the whole algorithm crisp, but then he would have narrowed the window of recognition to a slit. He would have had to state exactly how much one axis exceeded the other and thus accept some elongated shapes and discard others a tad shy of the cutoff. He would also have banned all characters whose axes were not exactly perpendicular. As it is, this method erects a large net to catch ovallike shapes. It allows fuzzy input.

It also tells how ovallike they are. It is a crucial point.

An oval is a letter in the Roman alphabet: the capital O. If the machine had similar algorithms for every other letter, it could read a character, determine the membership for all 26 algorithms, and pick the highest. If the shape's top membership was in V, the computer would declare it a V. The machine would decide that it is more a V than anything else.

It is very useful information. Crisp algorithms identify only ideal letters. Since no one cares about ideal letters, these algorithms have merely tossed pebbles at the great iron door of handwriting recognition. "The answer is never really black and white," says John Vieth, senior electrical engineer at NCR, which is investigating this problem. "It's never a definite call between a 3 and an 8. The decision is made on a sliding scale." Perfected, fuzzy algorithms could read odd fonts and extremely sloppy characters, the drunk-on-a-rollercoaster kind that tax our own ingenuity.

Handwriting recognition may seem an impressive if not revolutionary feat, but in Japan it is critical. The 3,000-plus written characters in daily use have long thwarted keyboards. Even the cleverest software has not created an easy-to-use keyboard for Japanese. As a result, the Japanese rely much more than Westerners on handwritten documents and faxes. A computer that grasped Japanese characters would go far to computerizing and streamlining their society.

On April 1, 1990, Sony announced a significant advance in this area: a slim portable computer that read handwriting. It was called the PalmTop PCT-500, a 2.8-pound device that opened like a book to reveal a screen on which the user writes with a stylus. It employed fuzzy logic to recognize not only the 26 letters of English, but 3,500 characters of Japanese, and it cost $1,320—a pittance. It worked by identifying over 300 stroke patterns and merging them into separate letters, a much more complex process than recognizing English characters.

Speech recognition has long been a grail of computerdom. A machine that recognized speech could react to oral commands and even take dictation, turning utterances at once into letters on the screen. Such a device would have major ramifications. It could expand the realm of computer users, downgrade the keyboard, and perhaps slash the number of secretarial jobs.

Yet speech recognition is much subtler and more difficult than handwriting. The obstacles seem preternatural. People instantly recognize most utterances, regardless of the speed of delivery or pitch of voice. We can thus understand the speech of many diverse individuals under an array of conditions. Computers, housed in their sensory straitjacket, are not so fortunate.

To date, most successful speech recognizers understand just one person. That individual must train the device word-by-word until it becomes accustomed to its master's voice. Moreover, these machines normally identify only isolated terms, not connected speech. And their range is narrow. Experts grant them a large vocabulary if they can distinguish 1,000 words.

Some speech recognizers can grasp anyone's voice, but they pose greater challenges. These devices have a large vocabulary if they can identify 20 words. They must overcome two hurdles: duration and pitch. Length of sound can vary widely even in one individual, according to context, emphasis, tension level, or external circumstances. A person in a hurry, for instance, will use shorter vowels. Pitch varies across individuals and is not a simple one-peak curve. Rather, people speak with many pitches at once, which make the voice richer and more interesting but give headaches to engineers.

Researchers at Ricoh, following a fuzzy model similar to Zadeh's, have developed an all-speaker machine that recognizes 120 words.[12] It is also much smaller than rival devices. It scored 93.2 percent accuracy in Japanese, 92.8 percent in English, and 93.7 percent in German. Michio Sugeno used one of these devices with his self-parking car. Of course, a 120-word vocabulary remains minuscule, and Ricoh is still working on this machine.

FUZZINESS AND PHONEMES

How do people recognize patterns? No one really knows, but fuzzy logic may play some role in it. For instance, consider speech sounds. Linguists like Roman Jakobson believed that we perceive each phoneme—each k, b, p—categorically. "The traditional dogma, for three or four decades, was that we don't notice differences within a category, only between them," says Dominic Massaro, a professor of psychology at UC Santa Cruz. "That says speech is a unique domain." For instance, *p* and *b* are almost the same sound in English. In both cases, we bring the lips together for a moment to hold up air, then release it. But with a *b*, we add a hum from the larynx; with a *p*, we don't. All *b*'s sound identical to us, said linguists, and so do all *p*'s.

In 1978 Massaro and Gregg Oden, then both psychology professors at the University of Wisconsin, declared that we recognize phonemes with a fuzzy logic model.[13] "We've shown that people do have continuous information in speech," Massaro says. "They can convey the extent to which a phoneme is a member of a category."

Their model of phoneme recognition closely parallels Zadeh's algorithm. It has three stages:

1. *Feature evaluation.* First, we assess how much the building blocks of a sound are present. For instance, we hear both parts of the *b*—lip stop and expulsive hum—to different degrees. The mind determines how much.
2. *Merger.* The mind then fuses these values into an overall degree for the phoneme. Oden and Massaro used a fuzzy formula here. For instance, the amount for *b* is the degree of lip stop times that of hum.
3. *Matching.* Finally, the mind compares the sound to a set of models, determining how well it fits each. It roams the various prototypes, selects the best match, and flags it as the phoneme.

Of course, they observed, the brain also relies on context. If a teacher says, "Tomorrow there will be a sbelling test," we will likely hear the *b* as a *p*, because only "spelling test" makes sense. But they also stated that the available context choices were likely fuzzy as well, and played a role in classifying the pattern.

Massaro and Oden would go on to apply this approach successfully to many other areas of perception, including letter and word recognition in reading, syllable and word recognition in speech, perception of visual depth, and categorization of visual objects. In each area, there were key fuzzy features, and the brain, it seemed, assessed them, compared the assessments with patterns in memory, and selected the best fit.

Yet as natural as fuzzy logic is at recognizing patterns, it still cannot learn them, as the brain does. That feat would require a new kind of computer, indeed, a new wave in machine intelligence: the neural network.

12

WEBS OF COGNITION

■

The machine does not isolate
man from the great problems of
nature but plunges him more
deeply into them.

—ANTOINE DE
SAINT-EXUPÉRY

The principle of the mind is the
Great Ultimate.

—ZHU XI

■

It is early 1991. Bart Kosko leans back behind the desk in his office at USC, contemplating the Patriot missiles bringing down Iraqi SCUDs over Israel and Saudi Arabia. The day before, one glanced off a SCUD above Tel Aviv, deflecting it rather than destroying it. The SCUD hit the city, injuring over a hundred and killing three. Kosko says a more accurate target-tracker based on fuzzy logic, which he has devised, might have stopped the SCUD in midair and prevented this suffering.

His target-tracker is an adaptive fuzzy system, a hybrid of fuzzy and neural networks which may rejuvenate AI. It unites the best of fuzzy logic with neural networks, computers based roughly on the brain.

Neural networks are among the most tantalizing advances in recent technology. They resemble the human brain—though remotely, like a paper plane resembles a jet. It is a big step forward, though, since a digital computer resembles the brain like a tossed rock resembles a jet.

Neural networks "learn." That is, they use experience to become better and better at classifying, that basic feat of thought. Exposed to enough examples, they can generalize to others they haven't seen. For instance, if a character-recognizing neural net sees sufficient A's, it can come to recognize A's in fonts it has never encountered. They can also detect explosives at airports, bad risks for mortgage loans, and words in human speech.

Moreover, neural networks can spot patterns no one knew existed. For instance, Chase Manhattan Bank used a neural network to help it reduce costs from stolen credit cards. It examined an array of information about thefts and discovered, intriguingly, that the most dubious sales were for women's shoes between $40 and $80.

The first commercial neural net product debuted in June 1992. Based on a neural net chip developed by Federico Faggin and Carver Mead, it appeared in a check scanner, a device that scans the code at the bottom of checks and electronically consults the bank about it. It could speed up buying by check.

Neural nets differ dramatically from the machines that now sit on the desktop. Traditional computers feed all information through a single point for processing. Neural nets process it everywhere and simultaneously. Digital computers store information at numbered addresses in memory. Neural nets store it throughout the system, at locales reached through content. Digital computers are brittle and can fail with one damaged part. Neural nets are robust and degrade slowly and gracefully, like Yamakawa's inverted pendulum.

Spanish neuroanatomist Santiago Ramón y Cajal (1852–1934) first suggested the central idea behind neural nets in 1893: The brain holds memories as patterns of linked neurons. In 1943, neurologist Warren McCulloch and mathematician Walter Pitts proved that a machine based on this principle—a neural network—could perform any feat that a digital computer could.

In 1949, Canadian psychologist Donald Hebb expanded on Ramón y Cajal's idea. We store memories not in individual cells, he said, but in cell assemblies, in strings of cells. Every time electric current passes from one neuron to another, across the microscopic gaps called *synapses*, it forges a better connection and makes it easier for the current to pass the next time. A synapse is like a tollbooth at a bridge, but a special, friendly kind where the toll decreases the more often you cross. A driver among a delta of islands, an irregular mosaic, would tend to take the route with the least toll. Current flows that way in the brain, Hebb said. Any oft-followed path would constitute a cell assembly, a memory, so the brain would have assemblies for, say, *square, rabbit*, and *sea*. This scheme could work because the brain has between 10 and 100 billion neurons, and each can have 1,000 to 5,000 synapses.[1] The number of possible cell assemblies is colossal.

In the 1950s Frank Rosenblatt began building machines based on this notion. He called them *perceptrons*, and they are the forerunners of today's neural nets. They are complexes of "neurons," all tied together electrically. The connections have different weights, and the greater the weight, the easier to cross. Each connection amounts to an association. Like human associations, they grow stronger or weaker with use. When they reach 0, they are forgotten. When they reach 1, they occur automatically, like the link between bell and salivation in Pavlov's famous canine.

Neural nets learn mainly by adjusting the strength of their synapses. If we show a net a G and it responds with G, it has hit the bull's-eye and receives positive reinforcement, which strengthens the connections used. It is more apt to follow that path next time. The neural net is the embodiment of John Watson and B. F. Skinner's behaviorism, now fallen so far from glory. In effect, the network tries out many different patterns, receives rewards for nearing the solution, alters the weights, and tries again, and again, and again.

In 1969, in a subsequently notorious episode, AI proponents Marvin Minsky and Seymour Papert of MIT published a proof that Rosenblatt's perceptrons, as then constituted, had stark limitations. The logic was unassailable, but most readers assumed it covered all versions of the device rather than just the elementary ones then employed. Perceptrons seemed a cul-de-sac. Interest chilled, and research shifted almost wholly to AI based on symbol processing. By the 1980s, when symbol-based AI was flopping about on the deck, researchers discovered that, with a few modifications, neural networks could sidestep Minsky and Papert's objections. The two have since incurred odium as suppressors of the field.

Do neural nets model the brain? The idea has provocative explanatory power. For instance, it makes very clear why people can identify patterns so much more easily than they articulate them. The patterns form first, gradually. We don't sense them until they reach a certain strength, and only much later, perhaps, do we translate them into words. Even then, the translation may be misleading. The neural net model also feeds ammunition to exponents of "intuition" as a more basic feature of the brain than symbol-oriented reason. Intuition, hunches, are our sense of the unarticulated patterns themselves. As makers of intelligence tests advise, guessing is usually a good idea if one has a hunch about a question. It can tap an incipient truth.

Regardless of such conjectures, as machines, neural nets differ fundamentally from brains. Most obviously, they have no emotions, creativity, or private thoughts. They lack the fabulous intricacy of gray matter. Their neurons are much simpler than real ones, which come in many varieties and relay information in numerous ways. Finally, the brain can devise rules from experience. Neural nets merely get better and better at recognizing patterns. As AI advocates have pointed out, they can't derive structured rules, and this inability not only sets them apart from brains, but hobbles them as rivals to AI. Fuzzy systems would solve this problem.

STRUCTURE AND NUMBER

By the dawn of the 1990s, Bart Kosko had emerged as a leading figure in both fuzzy logic and neural nets. Though he was only an assistant professor, his 1991 work *Neural Networks and Fuzzy Systems* had sold out on publication and become a best-selling text. Beyond reformulating the theory of fuzziness, he has helped combine it with neural networks to create a powerful new hybrid.

Kosko contrasts AI and neural nets with fuzzy systems. He notes that traditional AI is one-dimensional and its basic unit is the symbol, a large, awkward item. Yet it also has structure, that is, rules, and rules are priceless shortcuts. "The good news is you can represent structure in knowledge," he says. "The bad news is you can't do much with it because it's symbolically represented. In other words, you cannot take the derivative of a symbol. So the entire framework of mathematics and most of the hardware techniques for making chips are not available to you in AI."

Neural nets are just the opposite. They have the advantage of number. "You can prove theorems and you can build chips. The problem is that neural nets are unstructured." They cannot handle rules. For instance, a traffic-control system attempts to spur the flow of cars through city streets. What should it do if traffic grows heavier in one direction? The answer is pure common sense: Keep the green lights on longer. Unfortunately, one cannot just tell a neural network to do that. "You have to give it lots and lots and lots of examples," he says, "and then *maybe* it'll learn."

Moreover, like brains, neural nets do not have indelible memory. They are volatile. "You can't be sure it won't forget it when it learns new things. And that's the problem with neural networks, and that's why you don't have neural network devices in the office, in the factory."

The best of both worlds, he says, is the adaptive fuzzy system. (It is adaptive because it changes over time, to learn.) It has structure like traditional AI, so it can use rules. It also allows the pure math of neural nets, and thus chips and learning.

BRAIN SUCKING

Judea Pearl says fuzzy systems seem no different to him than ad hoc systems, devices engineers have long cobbled together to fit the circumstances. "I think Pearl is right from what's he's seen," says Kosko. "They are *beautifully* ad hoc systems, in the sense that the deep principle of intelligence is that we don't know the input–output transformation. The expert can't articulate it, though he can act as a function. And that's the point."

AI systems, he says, require us to fully specify all the rules, which can be nearly impossible. We often just don't know them. Neural networks, on the other hand, learn by example and only by example. But adaptive fuzzy systems can take the inputs and outputs of neural nets and express the relation in fuzzy rules. "It's basically detective work at the mathematics level. We want to estimate the rules just from the data, without the guess of a math model." The neural net behaves, and the fuzzy system divines the laws of that behavior.

In fact, deriving the rules has long bedeviled makers of fuzzy systems. As Zadeh notes, "The way it's been handled in the past—and it's still handled that way—is to build a system and see if that works. If it doesn't work, you begin to tinker with things."

The quest to automate this process goes back to a paper by Mamdani and an associate in 1977. In 1984, Tomohiro Takagi and Michio Sugeno published an article which discussed, for the first time in detailed and useful fashion, how to obtain rules by observation. As Zadeh notes, "Let's consider the problem of parking a car. There are two approaches. One is through introspection. That is, you sit down, you analyze the way you park the car and you come up with the rules. And the other is based on observation of somebody else's approach, not necessarily your own. And that is an important issue. This whole symbiosis of fuzzy logic and neural networks has to do with this problem basically: the induction of rules from observation."

Kosko has devised a learning method for machines that takes data straight from the outside world. He calls it differential competitive learning. "You ask Itzhak Perlman how he plays the violin and you'll get an answer. But you can't use that answer to replicate his performance. He doesn't give you an equation for mapping inputs on the page to tonal frequencies coming out of the violin. He doesn't know that and neither do you. Well, neither does a fuzzy system and neither does a neural system."

But observation of violin-playing can lead to rules. So can, say, trapshooting. "We take you and you sit down and try to put the crosshairs on the target," he says. "You take your best shot and what you're doing for us is generating trajectory data. And while you're doing that, it's cranking through this competitive learning algorithm and very quickly we're generating boxes of rules. It's called brain sucking."

The system can then take over and use the rules it has inferred. "So this is a real step toward automation at the intelligence level," Kosko says. "You yourself could not articulate most of the rules. In time you could, but we don't have to do that anymore. You don't need to talk. You just have to behave. That's why artificial intelligence has collapsed, because you can't articulate those fine rules. The problem with most fuzzy systems to date is that they require so much to articulate the rules, it takes a long time just to get a few."

THE FUZZY KALMAN FILTER

Kosko did not select his trapshoot example on whim. It relates to the Kalman filter, which he calls "the single most powerful, most popular algorithm and system of modern engineering." This Bayesian technique helped put men on the moon and bring them back, and engineers use it for most navigation as well as analysis of the bloodstream and other tasks. "If you're trying to follow an airplane," he says, "and it goes behind a cloud, you still have to estimate where it is behind the cloud." The Kalman filter yields that estimate. It puts the crosshairs on a missile which may be moving at 2,000 miles per hour. The problem is twofold: The target is traveling very fast and the data are noisy.

Kosko has developed a fuzzy Kalman filter which he says exceeds the original. "We're taking the toughest benchmark, the Kalman filter, head-to-head, fair game, fair fight, doing the best you can with the Kalman filter and the best you can with fuzzy, and beating it," he says.

The fuzzy filter has two extra advantages: robustness and learning from life. Like other fuzzy systems, it is robust in the sense that injury causes its performance to decline gradually, not abruptly. If one reaches in and begins randomly erasing rules, the fuzzy filter performs "quite well" until about 50 percent of them are gone. The Kalman filter, in contrast, "degrades very quickly." Kosko also tested it by inserting a few foolish rules, such as "Always turn left," and "again you find that unless you really do a major lobotomy, it performs fairly well." In addition, "it uses in-flight experience to modify its logical structure."

The Kalman filter guided Patriot missiles against Iraqi SCUDs. It is the greater accuracy of his fuzzy Kalman filter, Kosko says, that could have saved lives in the Mideast.[2]

ORCA CALLS

A guest at a party is speaking a foreign language. Another person overhears it and identifies it as Arabic, without knowing a word of the tongue. "How can we do that? How does it happen?" says Rod Taber, of the University of Alabama at Huntsville. "It's mysterious."

This kind of unarticulated pattern recognition had always fascinated Taber, and the tall, burly engineer eventually addressed an analogous problem with killer whales. He was working with General Dynamics in San Diego and became friends with marine biologists at the Hubbs Marine Research Institute. They had a variety of killer whale calls on tape, classified into types based on click, chirp, and whistle content. These types were dialects, and they showed the origin of each orca.

Scientists had tried and failed to identify these half-second sounds with machines. Taber set out to solve the problem. He borrowed orca calls from Norway, Canada, Antarctica, Iceland, and Alaska, as well as recordings of other underwater noise, such as ships, torpedo launches, helicopters, earthquakes, and fish signals. He then tried to build a neural computer to identify the killer whale dialects. "It worked terribly," he says. "You might as well have thrown dice. We figured there had to be a way of doing this, but something big was missing. So we jumped in with fuzzy logic."

He and his group invented a fuzzy neuron, a new hardware device, and in 1987 gave a demonstration to the Navy. The Navy handed Taber one of its own orca tapes and asked him where it came from. "It was the funniest darn thing," he recalls. "According to fuzzy logic, it had about a 75 percent chance of being from the southern end of Alaska. Nothing came out over 75 percent. I said, 'The machine doesn't recognize it per se. But if we had to bet, it would be southern Alaska.'" In fact, the Navy had recorded it 20 miles south of Alaska.

The Navy reacted in an interesting way. "They sat me down and gave me a talking to," he says. "They said, 'This is the way you detect submarines. The Russians could use it to detect our subs more efficiently.'" Chastened, he cut back on his publications in this field. "I still talk about it, but I tone it down and try to avoid touchy issues."

NEURO-FUZZY IN JAPAN

The Japanese have also been quick to exploit this technology. For instance, at Mitsubishi's Industrial Electrical and Systems Development Laboratory, a cluster of deceptively nondescript buildings near Osaka, Atsushi Morita and his co-workers devised a mixture of fuzziness and neural networks that possessed strengths of both.

Morita is a bright, cheerful man of 38 who grew up in Nara, the ancient temple city that was Japan's first capital. He first heard of fuzzy logic through Masaki Togai, whom he met at a conference in December 1985. At first, he failed to fully grasp it. "I thought it was for understanding language, not for control. In 1986, it was something new, but we didn't know how important or useful it might be." However, he employed it successfully in an electrical discharge machine, and by 1990 had developed a fuzzy neural network model.[3]

It worked fairly simply. Most neural nets start with their connections weighted the same throughout, that is, with no knowledge. Morita's model began with rules from experts. It sprang at once into being, like Pantagruel, then refined its rules even more. This tactic slashed learning time dramatically. Moreover, as it improves the rules, scientists can see how they change. The rules are not lost in the black box, but shift before one's eyes. Overall, these systems telescoped development and side-stepped much of the difficulty of deriving rules.

In nearby Osaka, Matsushita's Central Research Lab was using neural nets with fuzziness in commercial appliances. Manager Noboru Wakami had also encountered problems in determining the rules. The neural net proceeded by trial and error, an arduous exercise. "So we researched ways of automatically tuning the membership functions," he says. "For example, we combined neural networks with fuzzy logic. We decided the membership functions using a neural network learning algorithm."

Matsushita describes neuro-fuzzy logic as "fuzzy logic designed by neural networks." The company notes that it required some time to determine the fuzzy IF–THEN rules. Moreover, fuzzy logic by itself could only consider a limited number of factors simultaneously. Of course, people have this problem too. If a driver on an icy road at night begins to skid, she must contemplate an array of weighted factors at once—perhaps too many. Bad decisions ensue.

Matsushita found that, by taking data from experts and feeding it into neural networks, the nets could generate fuzzy rules automatically, 45 times faster than previous neural networks. Moreover, the machines could consider many more factors.

The first Matsushita products did not profit from this approach. But in February 1991, the company introduced its first neuro-fuzzy washing machine. Its original fuzzy washer adjusted for three variables: kind of dirt, amount of dirt, and load. The neuro-fuzzy machine could also handle type of detergent, quality of clothes, hardness of water, and other factors. It chose the most appropriate from among 3,800 different patterns of operation. The company soon also rolled out a neuro-fuzzy vacuum cleaner and rice cooker, and planned to place the technology in many other products.

Other, more advanced devices are blooming in the East.[4] Japanese researchers are working on a fuzzy-neural character-recognizer that can identify letters in extreme noise, such as where ink spills over half the letter. While neural nets alone cannot achieve this feat, neural nets enhanced with fuzzy logic can. Another project allows robots to cut unknown metal surfaces. In addition, Chinese researchers are developing neuro-fuzzy systems for diagnosing silicosis and predicting change in foreign exchange rates.

FUZZY COGNITIVE MAPS

Meanwhile, Bart Kosko was working on a new kind of decision system: the fuzzy cognitive map, or FCM (which he pronounces with internal schwas). FCMs are networks rather than one-way trees. They model situations by their classes and the links between them, and ultimately form a web of interlocking causes and their strengths.

"I claim I can take any article and translate it into a fuzzy cognitive map," he says. For instance, he derived the map in Figure 18 from an economist's analysis of the complex political situation in South Africa in the late 1980s. The question was whether the United States should divest. The map has nine variables, such as mining, black tribal unity, and white racist radicalism, and an increase in one can create an

Fuzzy Cognitive Map

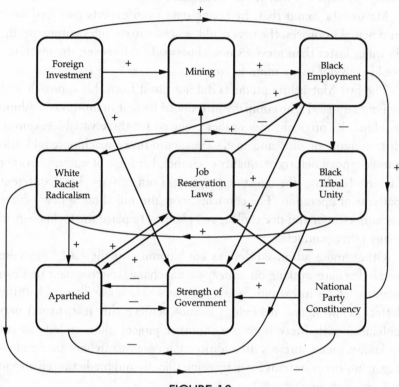

FIGURE 18

intricate domino effect, causing many others to rise or fall. (Figure 18 uses simple plus and minus signs to show increase or decrease, but a true FCM would have numerical values.) For instance, it shows how a rise in black employment would ultimately affect apartheid. The FCM is a dynamic system. It seeks equilibrium, and the final equilibrium is the inference.

"In fact," Kosko claims, "the article does nothing but flesh out the cognitive map. So the prediction is that soon you'll see political articles on the op-ed page, and then they'll have an appendix which is the cognitive map. A few years from now, it will be just the reverse. The op-ed page will have the cognitive map and the words will be the appendix."

Fuzzy cognitive maps reduce political analysis to a matter of identifying variables, the links between them, and the strength of the links. If the situation alters, one can easily adjust these links. The FCM thus responds at once to feedback.

Moreover, he expects analysts will be able to work easily with the maps, as they have with other fuzzy devices, because they can comprehend them. "They don't have to understand mathematics. They can just common-sense argue it. We can go through it link by link." It is, he says, a picture.

Kosko's FCM of South Africa proved controversial, perhaps partly because he adapted it from an article by Walter Williams, a black conservative. When he published it, the editors insisted he remove the labels like "Foreign Investment" from the nodes, and Kosko, angered, put "Censored" in their place.

Yet unlike conventional expert systems, FCMs can be easily combined, so they need not reflect the sole judgment of a Walter Williams. "We can take any number of expert opinions, the more the better, and combine them into some unified FCM," he says. "In other words, the underlying knowledge you are trying to tease out of these experts, it comes out and it improves as you increase the number of experts." Moreover, one can fix the weight given each expert, so the opinion of a minor figure does not equal that of a high authority. The map also deftly handles contradictory assessments of the same relationships. "Fuzziness doesn't run away from contradictions. It's built on them," Kosko says.

"I foresee a day when you can take every document that's ever been written, beginning with the Sumerians, take all the technical journals and all the books, the world's accumulated knowledge—wouldn't you like to see what that FCM looks like? And the beauty of it is that once you've done that, once you've grown this immense FCM in the sky, all you've really done is initiated the real FCM. Because now you're going to use techniques drawing straight from data. So for succeeding eons, this map will be growing and evolving in a very structured way, in a way that no single mind can perhaps understand, and it will make very terse predictions about states of affairs."

Such a behemoth authority would generate social problems. "At first it would scare the hell out of people," Kosko says. "That's why everyone would have one." But Kosko believes that the history of science shows that "can implies ought." He likens FCMs to early computers themselves, which disturbed people until everyone had them on their desks. "This sort of stuff will be available to everyone very soon."

Kosko adds that mammoth FCMs could clarify arguments and thus better pinpoint disagreements. "You could build a Republican map and a Democratic map, and you can argue the links. Most of the links will be the same, except some of the relations between policy variables won't be there." One could make predictions on the basis of both "bushes," see which come true, and plunge into the faulty bush to fix the "twigs." Where the two sides disagree, other twigs would reveal the source of the conflict, and one could narrow the dispute to the real issue.

"Earlier attempts to do things like this fell down," he says, "because they used decision trees or they went too far and tried to write down equations, which I don't think is tenable. But you can argue factual cases much more easily than value cases. If you can bring that down and get all the links to agree in [positive or negative] sign, if not in magnitude, that's a great achievement."

In any case, he says, "The good news is: The bigger the map, the more the nodes and the more connections between them, the less any given change in magnitude seems to matter."

SEX ROBOTS AND CANCER ZAPPERS

With such powers, Kosko envisions fuzziness as radically altering our world, our sense of the role of computers. For instance, he says, the machines will routinely write novels. A computer might not pop out a thriller every day, but perhaps once a week or month. He predicts such a power within 20 years, and perhaps some form of it sooner.

Would any human being care to read a computer novel? "Well," he says, "I think that will take a certain amount of time," but he feels it is inevitable. It would follow the basic plot rule of Aristotle, with three acts in a ratio of 1/4, 1/2, and 1/4. "You can break this down more and more methodically, and that stuff is very easy to teach an AI system. So you'll get a nice crisp three-act structure with the conflict resolved, and everything you can articulate in the classroom will be housed in that book. Then locally it will be filled in. You want a Hemingway style, you get a Hemingway style. You pick your author."

Voice recognition will be commonplace, he says. The modern struggle with this task will seem ancient history, and current voice recognizers will take their rightful place in museums for citizens to chuckle over.

Automatic chauffeurs are coming too, he notes. Already, some companies are designing sonar devices to provide emergency braking on the freeways. If the car in front stops too quickly, it will pump the brakes. "The next step, and it's not in principle that hard to do, is to put down a layer of asphalt that's shot through with small emitters or receivers. In other words, smart roads and smart cars, and maps embedded in the system and a voice system to tell it where to go. It's all a matter of obstacle avoidance."

He also foresees sex robots. "I would look for machine intelligence products there, because so much of sex is just grossly sensory, and because of the rapid advances in ceramics, and because you really don't have great tactile differentiation ability." He says such a fuzzy neural system is not farfetched at all. "It can't just be a Barbie doll. It has to appear to have an infinite repertoire of behavior," he says. "Given any input, it will generate an output and those outputs will have great variety." The market would be vast, he thinks.

"Just imagine if someone in Kyoto comes up with flesh that, if you did a Turing test, people said, 'Yes, that's flesh,' and you put in a little water and warm it up and so forth, and it makes all the sounds. The technology is not that hard. I think also if AIDS persists, and it could go on and on like the plague went on for centuries, then you'd be forced toward sex substitutes." Such machines, he thinks, might profoundly affect marriage and the nature of the family.

The potential of fuzziness extends even farther. Kosko dismisses MYCIN and similar medical diagnostic systems as crude. "Forget that business," he says. "There are many other ways to do that."

For instance, Eric Drexler has speculated about a new domain he calls nanotechnology, the technology of machines the size of molecules. It could yield a bouquet of wonders, such as infinitesimal devices that automatically convert energy into food or roam the bloodstream in search of disease. This field is taking shape much faster than most scientists expected, and in 1990 researchers stacked 35 xenon atoms to spell IBM. Such micromanipulation will be critical to build these tiny machines.

"Real solid progress has been made," Kosko says. "But the problem with Drexler's concept was that he had this undefined idea in it: AI engineering," Kosko says. "In fact, he was really talking about neural or fuzzy or a mixture of both." Kosko thinks no one could ever build a large enough AI expert system, with its ponderous load of rules, to work efficiently at the molecular level. Yet the devices require machine intelligence. For instance, a molecular virus hunter would need to track fast-moving targets. Once it learns that trick, he says, it demands very little computational power. "You don't need a chip to do it."

The machine would also have to make judgment calls, which require more. "In other words, the idea of a cancer cell or a healthy cell—it can't call all these in advance. It'd have to have a neural-like system to recognize categories or patterns, not just specific instances. So I think that's the key part."

The ultimate promise of nanotech is the dream of ages: immortality. "The minute you start taking a quantitative view of health at the molecular level, there's really no reason you can't keep those molecules healthy and definitely slow the aging process down to nil." Individuals cryonically suspended—and Kosko believes strongly in cryonics—might then be brought back to enjoy lasting youth.

These visions have a sparkling appeal. But will they come true, or will they harden into garish embarrassments like the predictions for AI and neural networks? Overhype is a real danger to nascent technology. It creates expectations which go unattained, and ultimately infects a field with discouragement. Kosko may or may not be prophetic, but we can begin to form an opinion by peering into the labs where engineers are creating tomorrow's fuzzy products.

13

INSIDE THE JAPANESE LABS

■

Technologies considered "old news" in Tokyo, such as super-fast trains and biosensors, are viewed as "science fiction" by my colleagues in Silicon Valley, who seem totally unaware of Japanese developments in their own fields.

—SHERIDAN TATSUNO

I am convinced that the societies that master the new sciences of complexity and can convert that knowledge into new products and forms of social organization will become the cultural, economic, and military superpowers of the next century.

—HEINZ PAGELS

■

One morning in April 1988, Michio Sugeno's phone jangled. When he picked it up, he found himself talking to Taizo Nishikawa and Masataka Nakano of MITI. Sugeno knew their names very well. In 1981, Nakano had spearheaded the effort to establish Japan's Fifth Generation project, or ICOT, a 10-year $850 million exploration of AI, and Nishikawa now held Nakano's former post. Sugeno half-expected their call, since he knew MITI had become interested in fuzzy logic.

They gave him important news. MITI wanted to create a new laboratory to investigate fuzzy logic. Would he like to put a group together and draw up a plan? Sugeno did not have to go into seclusion to deliberate.

The Japanese government funds a wide array of programs in high-tech—protein engineering, 3D integrated circuits, high-performance ceramics, advanced robotics, telecommunications, high-definition TV, optical computers, and perhaps 20 or 30 more.[1] These labs are not blue-sky projects, but highly practical endeavors—each a little Los Alamos in commerce. Big companies finance them, and not from idle curiosity. They reflect the profound importance Japan attaches to technology, the key to its success in the modern world.

Not all of these projects have succeeded. For instance, the much-ballyhooed ICOT never yielded the breakthroughs some foresaw, and its failure may have hastened Japan's shift to fuzzy logic. Yet ICOT was simply one more aspect of Japan's trawl for promising innovations, an approach that minimizes the chance it will slight good technology.

In May 1988, Sugeno's group first met with officials from MITI as well as from companies and other national laboratories. In June, Sugeno tapped a bright former student of his named Tomohiro Takagi, who had been out of fuzzy logic for five years, to help draft the proposal. Takagi had made key contributions to fuzziness and had worked with Zadeh in 1983. To qualify him formally for the project, Sugeno also secured a post for him at Matsushita.

MITI expected that seven to ten corporations would sponsor the project, a typical number. Sugeno, however, felt fuzzy logic had broader appeal and set about beckoning more companies to join, eventually enlisting 49. They comprised a who's who of Japanese technology—Toshiba, Nissan Motor, Matsushita Electric, Fujitsu, Nippon Steel, Hitachi, Sony, Canon, Minolta, Mazda, Sharp, Olympus, Toyota, Omron, Konica, Honda, and many more. "MITI was a little bit surprised," he says.

The group raced to meet a three-month deadline at the end of August, by which time the press was treating the project as a fait accompli. At one meeting, the name Laboratory for International Fuzzy Engineering Research suddenly popped into Sugeno's mind. "I was sitting next to Mr. Nishikawa and I wrote it on a piece of paper—LIFE—and showed it to him and he agreed."

The group was seeking a special grant established in the late 1980s to develop technology that would succeed in the United States, such as superconductivity and opto-electronics. "We supposed there might be some fight between United States and Japan, because the United States changed its policies," Sugeno says, referring to the 1986 U.S. trade restrictions on Japanese DRAM memory chips. "The aim was to overcome Japan or something like that. So therefore MITI thought we had to prepare."

Budgeted at ¥2.5 billion (about $20 million), their proposal went before the reviewing committee in October. This panel consisted of some ten people, including professors from ICOT, a project known as antifuzzy. Sugeno and his team were confident they would get the full budget.

To their astonishment, in December the committee rejected the plan, citing three drawbacks: Computer scientists did not appreciate fuzzy logic, it was unpopular in the United States, and fuzzy control worked only with small systems. "But it's not true of course," Sugeno says. "We already had the Sendai subway and most applications of fuzzy control at that time were for complex systems. Not for small systems."

The veto angered Nishikawa. MITI had brought many of these scholars into the national limelight through ICOT, and now they had publicly embarrassed it. So Nishikawa offered Sugeno a grant of ¥5 billion, twice as much. The team resubmitted the proposal in January 1989 and won approval at the end of February—in just one month. "So ironically I can say we succeeded because of that committee," Sugeno says.

LIFE was to last six years. The plan called for a budget in two halves:

¥2.5 billion from the government and the same amount from the corporate sponsors—a total of some $40 million for the duration. Sponsors fell into two groups: A and B. Group A firms paid about ¥10 million ($77,000) per year and could send researchers to LIFE, while group B firms paid about ¥5 million and could not.

Sugeno suggested to MITI that his mentor Toshiro Terano act as director of LIFE. Sugeno himself became "leading advisor," a title that belied his true influence. (Six months later, LIFE gained a second advisor: Lotfi Zadeh.) Tomohiro Takagi came aboard as vice chairman of the fuzzy information processing laboratory.

THE EVOLUTION OF LIFE

LIFE opened its doors on March 15, 1989, in the gleaming Siber Hegner building in a chic part of downtown Yokohama, two blocks from the town's baseball stadium. It was and is the world capital of fuzziness. With its broad corporate base, research staff of 27, and support staff of 18, it pursues the most advanced fuzzy agenda on earth.

Toshiro Terano is a distinguished, delightful gentleman who merges the best of paternoster and pixie. He chuckles often as he speaks, as though continually amused by the oddities of life, and quickly makes guests in his large corner office feel comfortable. "The motto of this laboratory is to establish human-friendly systems," he says. "Our idea is simply that the ordinary computer and human make good human friends. The computer can be used by any sort of person. For example, housewife, old person, young boys, girls can utilize it easily, like a friend."

Though the presidents of some of the world's largest corporations sit on LIFE's board of directors, Terano cautions against interpreting this fact Western-style. "In Japanese organizations, the president is not almighty," he says, referring to the ubiquitous practice of *nemawashi* (literally "binding the roots"), or decision by consensus. "Many, many people are discussing it, many managers and many directors are discussing this policy. But most of the heads of companies agree with it. So I think this is a very good idea," he laughs.

As the project progressed, it came to absorb Terano the engineer, and he more and more immersed himself in research. By mid-1991,

Tomohiro Takagi had assumed the main duties of administration. It is a tricky job, like herding cats. As a consortium, LIFE enjoys prestige, prominence, and funding. However, it must also satisfy its member companies, who often tug in their own directions.

For instance, some have treated LIFE as a postgraduate school. They send researchers with little knowledge of fuzziness who spend their brief tenure assimilating it, then return to the firm to teach other employees. These students have been a drag on LIFE's ambitious goals. The practice is common in Japan and good for it, but not for LIFE. Indeed, says Takagi, at first Terano was not sure the project would thrive at all, because of its 49-headed nature. Now, Takagi estimates that 80 percent of its projects will succeed.

A second problem soon appeared, one of structure. When Sugeno designed LIFE in 1988, he plotted out three main research areas: fuzzy control, fuzzy intellectual information processing, and fuzzy computer. It was a broad program, and LIFE spent the first two years shaking it down.

"We have to be a pilot for research and development," Takagi says. LIFE seeks to bridge the gap between basic research and commercial products. It strives to build working models from theory, as Mamdani did with his steam engine. Once it completes such protoproducts, companies finish them, and their cooperation turns into ferocious competition.

However, both fuzzy control and the fuzzy computer strayed from this protoproduct goal. Fuzzy control proliferated so rapidly in Japan that LIFE soon had little contribution to make and risked duplicating efforts at corporate labs. On the other hand, the fuzzy language computer lay so far in the future that it required not just research, but basic conceptualization. It held little near-term promise, and few sponsors were interested in it.

Thus, in September 1990, LIFE reorganized. As Takagi says, "First we asked: What is the main goal and what is the main duty for us? It is the achievement of intelligence using fuzzy theory." Hence, LIFE shunted fuzzy control and the fuzzy computer aside, and placed the spotlight on fuzzy intellectual information processing. But the fuzzy computer is Sugeno's great goal and he could not abandon it, so he started a working group on fuzzy computing in parallel with LIFE.

Like most MITI research consortiums, LIFE has a time limit, in this case six years. However, these bounds are elastic. "There is no consortium finished," Sugeno says. "So therefore the LIFE project will be extended, or we shall be able to start another laboratory." For instance, MITI has now launched a Sixth Generation project, which will combine neural networks and conventional AI, but not fuzzy. "But MITI is interested in fuzzy logic," Sugeno adds, "so we are quite certain we will be able to start another project after LIFE."

THE PROJECTS OF LIFE

LIFE is researching along a wide front, one that includes decision support, language and image understanding, intelligent robots, and fuzzy neural networks. At this writing, many projects remain under wraps, but the details of some, especially those emerging as the more successful, are provocative.

For instance, LIFE is working on a decision-support system for foreign exchange trading, an expert system somewhat like Chizuko Yasunobu's bond trading program. Called FOREX, it incorporates the effect of political events as well as economic ones. It will predict the exchange rate of the yen against the dollar using not only market conditions, but also news about national political policy and statements from trade leaders.

Takagi says this challenge stymies the usual AI methods. "If we can do this, it will prove that fuzzy theory is an essential technology to resolve such complex problems." By mid-1991, he felt this project had an excellent chance of success.

LIFE is also investigating fuzzy management of power plants, with systems that can understand events occurring in the plant. Such systems could help prevent accidents. For instance, LIFE researchers note that warning signs normally foreshadow an accident. If a specialist spots them, she can avert the mishap. But people nod. This model would monitor the plant constantly and without flagging, and would know enough to identify all danger signals. Like a fuzzy bond trader, it would highlight patterns people might ignore. In addition, fuzzy logic would capture the knowledge of the plant designer.

LIFE is also researching fuzzy control of a biological process: sake fermentation. Like cement kilning and water purification, fermentation takes a long time and frustrates feedback. This technique too has obvious economic applications.

As one might expect, LIFE is pursuing how to feed in, represent, use, and generate vague information. But it is also studying how to deal with background knowledge, which is crucial to grasping language.

Sugeno's self-parking car understood language to some extent. It grasped the assumptions behind "turn right" and carried the command out correctly. Such underpinnings—broadly, context—are critical for a person or machine to truly understand language. Yet they are wildly elusive. LIFE is seeking means of solving this problem.

For instance, LIFE is devising ways for the computer to retrieve past examples and modify them when reasoning. It also wants the device to recognize analogous situations and use more than one experience at a time. Moreover, LIFE would like the computer to employ the intended meaning of the speaker when searching for analogous examples. Divining that intent requires knowledge about the speaker himself.

LIFE is also investigating computers that grasp images as humans do. It is another precipitous wall, since vision is so complex. For instance, changes in light or viewpoint alter an image, yet we adjust so easily that we scarcely notice it. Machines do not. They lack the benefit of millions of years of evolution. Even a moderate level of image understanding would have commercial value. For instance, it might help robots on assembly lines pick up bolts instead of nuts.

One project involves developing a fuzzy system to see colors like people. Such a device might, for example, identify skin color. It could also summarize complex color patterns as "bluish" or "reddish," and use this information to separate, say, cans of soda or bottletops. Another project involves matching representations of human faces with images, which may also entail language to some extent.

LIFE is building a home robot, a friendly, roving device with an angled arm somewhat like an elephant's trunk that could perform household chores and other tasks. It would control itself and make intelligent decisions on its own, as people do. For instance, LIFE researchers have written an algorithm that approximates human avoid-ance behavior, allowing the robot to scoot around objects based on their perceived level of danger. It could acquire skills and knowledge, and evaluate large-scale situations based on sensory information and vague methods of estimation. It could also detect people or animals and tell them apart from furniture.

Takagi has devised the first practical implementation of Kosko's FCMs, which he calls "a very powerful framework to represent or execute knowledge of human intellectual processing." In Tokyo there is a large underground reservoir of wastewater that must be pumped out regularly. If a heavy rain begins to fall, the city wants to begin pumping at once, before inflow sets in, and needs to know how much to pump. It is an intricate problem, with many interacting factors, such as the level of the rivers. Like sake fermentation or cement kilning, the phenomenon also involves a time delay.

"We thought FCM seemed to be the best to describe this causal relation," Takagi says, and he devised an FCM to control the scheduling of pumps. Had he been working in private industry, he might have applied mathematical statistics to this problem, a much safer approach. The FCM was untried and risky, but LIFE exists to take such chances.

It worked. "So our model made by FCM has almost the same performance as a very precise model," Takagi says. Once he had created this technology, companies could take it up at once. Moreover, he and Sugeno then had a framework for modeling dynamic processes, and they took it straight to another daunting problem.

Michio Sugeno is now working on a much more ambitious "robot": a remote-control fuzzy helicopter. "It's very complicated," he says. Such a vehicle, regulated by satellite high above earth, could carry out dangerous missions, such as rescuing people on ships in a storm. "This unmanned helicopter is a kind of national project. It's sponsored by the Ministry of Transportation."

Any helicopter faces the problem of backrush. As it approaches the surface of water or land, it presses air harder and harder against it and generates a fiercer backrush of air. The pilot must adjust for it. Yet no crisp line separates the region with backrush from the region without.

To control the helicopter, Takagi proposed that they write IF–THEN rules in which the IF described a fuzzy location in space and the THEN was a mathematical formula. For instance: IF height is low, THEN [equation 1]. IF height is fairly high, THEN [equation 2].

Fuzzy-neural systems were vital to this project. Response time was critical, and a normal fuzzy system would have been too slow. "But a neural network has prewired links," he says, "so it executes much faster than fuzzy rules." Moreover, of course, it can learn and improve.

They fed in the fuzzy rules and tried it out in simulation, using a 3D graphic display on an Iris computer. "The first time, it was not stable, but after learning, it became stable very quickly," says Takagi. "We found that the system could control the helicopter very nicely."

Sugeno soon moved up to a real miniature helicopter, a flat circular device slightly larger than a paper plate, with four rotor blades, one at each corner. When a person operated this helicopter by remote control, it fluttered back and forth as though buffeted by wind. But the fuzzy controller held it almost stationary. The difference was striking. By now, Sugeno has devised about 100 control rules, with the goal of guiding the helicopter by oral instructions from the ground, and is working with a larger machine, three meters across.

People have operated helicopters by remote control for a long time. Indeed, U.S. servicemen in World War II controlled helicopter drones from shipboard. However, Sugeno's device is far more sophisticated. It will stabilize a helicopter even if it loses a rotor blade, a task not even human beings can perform. Bart Kosko calls this device the "premier example" of fuzziness in the world today.

THE LANGUAGE MACHINE

The boldest fuzzy project in the world today is also Sugeno's and aims at no less than a computer that will understand people—like HAL in Stanley Kubrick's *2001*, but without the distressing initiative. Instead of typing "OPEN FILE," one could say, "Let me see that report I was working on yesterday." It would then appear onscreen. The new computer might also speak sensibly, describe photos, and even summarize documents.

When Sugeno commenced his working group at LIFE in 1991, he worried that LIFE was addressing only fuzzy chips and fuzzy versions of conventional computer languages. He sharply refocused the inquiry. He wanted a fuzzy computer to represent "natural language," language as people speak it.

Language is one more goal the early AI theorists thought computers would quickly achieve, because we ourselves use it so easily. But in fact no one fully understands it. It is extremely complex, and it interacts with thinking and memory in deeply mysterious ways.

"That is our main problem in AI and computer science and also in control engineering," he says. "And there we have to deal with the uncertainty of information. AI people usually use probabilistic methods to deal with uncertainty. But the uncertainty involved in natural language is not probabilistic in nature but fuzzy. So therefore fuzzy logic is a key technology to deal with semantic information." The Fifth Generation computer dealt with only numerical and symbolic data, he notes with pleasure. "But from now on we have to deal with semantic data. Data have some meaning, and at present I don't know any other method than fuzzy logic."

Sugeno knows that fuzzy logic is not enough. Language remains far too elusive and contains many kinds of uncertainties. However, "one of the uncertainties is solved by using fuzzy logic. Anyway fuzzy logic is the most powerful tool to deal with natural language."

So his working group is laboring from scratch to build a fuzzy language machine. The group holds monthly meetings and plans to publish reports on its progress. Its first task is imagining what such a computer would even look like. How will it work? What will its functions be? Sugeno hoped to finish the conceptual blueprint in 1992.

After that, the group will move on to designing the hardware and software architecture, with the aim of producing some software by 1995. "At the moment we have only the fuzzy chip," Sugeno says. "But the fuzzy chip is not a computer. A computer is more than a chip. The most important thing is software." After the software, he plans to design the hardware architecture, so the software will reflect the basic concept and the hardware will reflect the software.

How would the hardware differ from current architectures? First, most computers today are binary. The new machine will be a hybrid, both binary and fuzzy. Second, computers generally have two main parts: processor and memory. The fuzzy machine will have three: processor, memory, and database.

The computer will mirror the left brain/right brain split in human beings. Neuroanatomists say the right brain tends to perform more intuitive tasks, while the left brain carries out the symbolic activities like math and language. The new computer will have both fuzzy and binary processors. "The fuzzy processor is close to the right side of the brain," Sugeno says. "The binary one is equal to the left side. And to combine them, we must have hybrid processor control." The fuzzy processor in turn will have three smaller processors: for fuzzy inference, fuzzy logic and fuzzy arithmetic. Likewise, memory in this computer will come in two types: fuzzy and binary.

But the crux will be the third part of the device, the database. "The first job of the fuzzy computer is to understand the meaning of a program," he says. "So it must have a database. It must have common sense." The database would organize memories, relate them to each other, and make them available for fast retrieval or consultation. If fully merged with processor and memory, it might supply context automatically, as the brain does. "Up to now, we just had part of this process," says Sugeno. "So maybe we have solved a couple of percent of the fuzzy computer. So 98 percent has not yet been solved."

By 1995, Sugeno would at least like to show what a fuzzy computer could do. "Maybe the best thing is simultaneous translation or making summary of a novel or article. Or, for example, you give it a photograph and the problem is to explain the meaning of it using natural language."

He wants to demonstrate a prototype in this century. If such a machine appears, it will so simplify computers that everyone will be able to use them. Well, perhaps not everyone. "We may have a very important issue," Sugeno says. "At the moment we . . . I'm not sure, but we may use Japanese for the fuzzy computer. That may cause a big problem between Japan and the United States."

But of course the computer should speak the language of its creators. Indeed, this is a project for which adjectives like *daring* and *ambitious* seem flaccid. If he and his working group succeed, come close, or even make it halfway, they will have cracked one of the great barriers in history.

OTHER LABS

One day in 1987, soon after the second IFSA conference, Sugeno visited Japan's Science and Technology Agency (STA) in the Toranomon area of Tokyo. STA is responsible for Tsukuba Science City, the famed "city of brains," a town of 12,000 researchers out among the rice paddies. STA actually has a larger budget than the better-known MITI and has funded inquiry in such exotic fields as bio-information transfer, bioholonics (biochips and biosensors), nanomechanisms, gallium arsenide crystals, ultrafine particles, and superbugs.

Sugeno asked agency officials to start a project on fuzzy logic. "STA had a special grant to promote new technologies and we made application in that year, but we failed," he says. Undeterred, he began gearing up to resubmit. In April 1988, Sugeno created a committee inside STA to promote fuzzy engineering. "I asked Professor Terano to chair that committee," he says. It consisted of about ten university professors and classified potential applications into six fields: (1) fuzzy theory, (2) fuzzy computers, (3) machine intelligence, (4) high-level man–machine interaction, (5) social science, and (6) complex physical phenomena.

Sugeno again offered the proposal and in February 1989 explained it and the promise of fuzzy logic itself to a screening committee. The proposal focused on fuzzy systems and included many areas, "from hardware to environment control, or even social programs." He also discussed the natural language computer.

Approval came through in March, a grant of ¥1 billion (about $7.6 million) for five years. Some of this money went to LIFE and another part to Sugeno's lab. The research group had 19 members.

Still another project commenced in April 1991. With the support of MITI, Sugeno and others established a fuzzy AI center, "a center to promote fuzzy engineering in Japan for fuzzy AI." Committee members who sabotaged the first LIFE proposal, Sugeno says, had proposed the original AI center. But it "didn't work at all, so therefore we put new life in it and we changed its name: Fuzzy AI Promotion Center." It stands apart from LIFE and has no set terminus. Sugeno says the center organizes seminars at companies and will create a working group on fuzzy control. The center publicizes fuzzy logic rather than investigating it. It communicates with firms, whereas the STA group carries out the actual research.

THE YAMAKAWA PROJECT

Takeshi Yamakawa was chafing at his bonds. He had built a fuzzy chip and written a popular book about it. But now, as a professor at the

National University, Kyushu Institute of Technology, he suffered legal restrictions on how he could use money for research. In addition, lab space at the university was not only scarce, but granted equally to each professor regardless of need. These restraints increasingly bothered him and he sought to sidestep them.

One way was to create an institute separate from the university, yet tied to it closely enough that he could make use of the bright students under him. "So I decided to establish the foundation," he says. It was not an easy task. By law, he needed a minimum of ¥100 million (about $770,000). So Yamakawa crammed all his classroom teaching into one day per week and began campaigning vigorously up and down Japan. Over a year and a half, he met 100 presidents of companies—hardware companies, software companies, companies of any kind which might need fuzzy logic—and asked them to donate money.

In the end, ten firms granted him a total of ¥150 million (about $1,150,000). He was in business. Finally, on March 15, 1990, exactly one year after LIFE, he received licenses from MITI and MESC (the Ministry of Education, Science, and Culture), and opened his Fuzzy Logic Systems Institute, or FLSI.

But though he finally had a shingle, he lacked a building to hang it on. Currently, FLSI is simply his own lab and a small office, so most research takes place at member companies. Hence, he at once set about planning a special headquarters for FLSI, a two-story structure to be completed in 1993. On the first floor, technicians will manufacture fuzzy and neural chips, and on the second floor, engineers will have their offices and meeting rooms. Currently, five staff members work at FLSI itself, but by then, he notes, "We can promote much more active research."

Soon after formation, in the summer of 1990, FLSI staged its International Conference on Fuzzy Logic and Neural Networks in Iizuka, a former coal-mining town near Yamakawa's base in Kumamoto. Lotfi Zadeh was there, and Yamakawa showed Bart Kosko the cave of Musashi, where the samurai lived and wrote. To climax the affair, Yamakawa held a dramatic international video discussion, transmitted by satellite between Iizuka and NASA in Houston. "It took 15 million yen [about $115,000]," he says. "I had to get the money within less than one month. It was hard for me, but two companies helped us."

Meanwhile, FLSI was working on four projects, two involving new types of fuzzy chip and two involving metallization in making chips.

FLSI first developed a new fuzzy controller chip and a fuzzy neuron chip. The latter fuses neural science and fuzziness, and he has used it to build a pattern recognizer. In July 1991, at the fourth annual IFSA Congress in Brussels, he demonstrated the system. It is portable and employs more than 20 fuzzy neuron chips to recognize the figures 0 through 9. Each neuron chip deciphers only one pattern, so to recognize ten figures requires at least ten fuzzy neuron chips. However, since a number like 4 can vary, the system assigns two or three chips for each figure.

Two other projects involved *metallization*, laying down fine amounts of metal on delicate surfaces. This process is vital for chip manufacturers and suppliers. Though uncompleted as yet, Yamakawa says, "The results are very good, because traditional technology facilitated only a low quality of metallization. But with fuzzy logic control, the metallization quality is very good."

FLSI has also pursued a separately funded program to boost the fuel efficiency of auto ignition. Starting a car consumes a great deal of gas—one reason city driving lowers the gas tank so much faster than freeway driving. Yamakawa developed an ignition system to maximize efficiency while using the least amount of gasoline. He created it as a breadboard (an assembly of separate transistors, capacitors, and other electronic parts), but was unable to fuse them together on a chip before the money ran out. "I wanted to promote the hardware system to one chip because it should be mounted in the engine," he says. "It is very, very suitable for engine control." He expected funding to resume by 1993.

Once such a chip appears, of course, the Japanese can quickly place it in a Honda or Nissan and use it to compete against American Fords and Chryslers. The U.S. cars won't have it, for as the Japanese probed the far reaches of fuzzy logic, Americans were just waking up to its existence.

14

AMERICAN UPSWING

■

Tomorrow is the most important
thing in life.

—JOHN WAYNE

The real concern I have is
whether they'll walk over the
white goods, the washing machines and refrigerators. If they
come in with low-energy fuzzy
machines, it could be a replay of
the auto business.

—SHERIDAN TATSUNO

■

In 1991, an Israeli ex-commando named Avi Nesher was in Hollywood planning to direct a sci-fi movie called *Hammerheads*. He signed on Bart Kosko as a technical adviser, and Kosko soon began rewriting the script, suffusing it with fuzzy technology and philosophy. The robotlike hammerheads became nano-tech superbeings, "fuzzy logical super-men." They possess fuzzy brain implants linked to their retinas, cochleas, olfactory bulbs, and other sense organs, and while today's systems program in English, Kosko says, they program themselves by pure desire. What they will, they become.

Hammerheads will be the world's first fuzzy movie. It is fantasy, of course, not sober prediction. The corporations of the world are not scurrying to hardwire brains and manufacture fuzzy ultra-Terminators. Yet it shows the rising interest in fuzzy logic, even beyond the realm of scientists.

Evidence for it abounds. George Klir notes an upsurge in enrollment in his fuzzy classes, and Kosko says the seminars he gives for private industry are now routinely oversubscribed. Lotfi Zadeh is so busy flying all over the world that he has vowed to cut back. Companies like Lockheed, Motorola, Martin Marietta, United Technologies, and NYNEX are cosponsoring fuzzy workshops, and many others are actively developing products. Some fuzzy theorists even worry that fights for funds and glory may now rend their once-tranquil discipline.

Yet the old biases remain alive. Speaking of the Sendai subway, William Kahan compared fuzzy logic to astrology and numerology, and said, "Whether or not to take an airplane flight based on astrology is comparable to deciding what kind of controller to use based on fuzzy calculus."

And funding barriers stand as well. In the fall of 1990, Kamran Parsaye tried to give a matching grant to Zadeh's students at UC Berkeley. The students wrote a proposal saying Parsaye and Rockwell International would equal the NSF grant, and sent it to NSF. "These are routinely approved," Parsaye said. "Yesterday I got a letter that Zadeh was turned down."

As recently as 1991, says Maria Zemankova of NSF, "Someone came into my office and said, 'Take this proposal. I don't fund fuzzy logic.' I said, 'This is in your area scientifically and just because it uses fuzzy logic to solve problems in your area, you can't say, "I don't fund fuzzy logic." ' " She shrugs. "He had no choice but to take it back, but I know what'll happen: He'll send it to reviewers who will kill it."

The United States still lags behind. It has dived into the race, but it still feels the shock of the chill, and its competitors are churning up the water far ahead.

DRUMS ON THE EBRO

Japan is not the only nation pulling away from the United States. Europe is moving forward as well, though less swiftly. In France, classical control engineers were a powerful force and hindered early fuzzy proponents. "There wasn't outright hostility," says Henri Prade, a computer scientist who founded a fuzzy logic institute in France. "It was more that when people were speaking about fuzzy sets, they were just laughing. They weren't taking time to look seriously at it."

By 1990, however, most of the hilarity had subsided. Academic researchers were winning grants and forging ties with industry, which is now intensely curious. "The phone is ringing every day," says Elie Sanchez, president of the Neural and Fuzzy Systems Institute in Marseille.

Hans Zimmermann says, "In other areas in which I was working, such as AI and business administration and statistics, people were very hostile. Until recently. Until the boom started in Germany in November 1990." Because of reports in the media, he notes, fuzzy logic has become a national cynosure. "Everybody's interested. It's fascinating in Germany now. There are companies, professors, researchers, students, even down to school teachers, secondary school teachers, all asking for demos. For example, I had a small presentation the week before last and the audience was 750 people."

Significantly, many in the European AI community did not dismiss fuzzy logic outright as their peers had in the United States. Professors in Spain, France, Germany, and Italy have advanced the field and cooperated on numerous projects. And Bayesians held much less power. "We didn't have to fight with these people," says Sanchez. "People like

Peter Cheeseman, he wrote some stupid things. He didn't understand fuzzy sets."

Moreover, fuzzy advocates often held prestigious posts. In Spain, the mathematician Enric Trillas, a pioneer fuzzy theorist, was president of the National Research Council, which administers 90 research labs throughout Spain in fields ranging from science to philosophy. Ramón López de Mántaras Badia, another fuzzy academic, is vice president of the Spanish AI Association. He now works at the Artificial Intelligence Research Institute in Blanes, 60 kilometers north of Barcelona on the beautiful Costa Brava. "I'm convinced that fuzzy logic is accepted at least in the AI community, which is not the case in the States," he says. In Europe, he adds, "The future looks quite good for fuzzy logic."

The European Community is funding a mammoth project called ESPRIT, which among other thrusts is probing uncertainty reasoning. The fuzzy logic research is taking place in a three-year subprogram called DRUMS (Diffusable Reasoning Uncertainty Management Systems). "It's a huge project, with around ten research groups involved," said López de Mántaras. Ebrahim Mamdani is also participating in DRUMS.

Private industry in Europe has become more active as well. Siemens, which *Fortune* ranked fifth among world electronics firms in 1990, is developing fuzzy systems to keep pace with its Japanese rivals. By the late 1980s, the manager of the automation group had noticed the Smidth kiln and he "asked us to see if we could do something similar," says engineer Werner Remmele. "There was lots of interest." In 1990, Siemens decided to apply fuzzy logic in more than ten divisions of the firm and went on to create its intelligent systems control lab.

Siemens has good reasons for pursuing fuzziness. Some 25 percent of its sales come from automation, that is, control systems. "We do automation techniques for everybody," says Remmele, and the Japanese success with fuzzy control arrested their attention. "We as market leaders had to study its potential, and we decided the potential of fuzzy control is big, so we had to develop this technology for our own applications."

Michael Reinfrank adds that Siemens must maintain its prestige by offering the best. "It is important for the rest of our products in conventional control," he says. For instance, he notes, fuzzy logic accounts for only about 10 percent of all control at the Smidth kiln. It's not a large amount, "but it is the core of the problem." A company might very well give an entire contract to Siemens just because it could handle the hardest challenges.

Reinfrank says it makes a big difference whether one talks to a control theorist or a control engineer about the merits of fuzziness. "A control theory person will say, 'We can solve any control problem in the world with our control methods.' That is true. You can control a cement kiln with conventional methods if you, for example, use 60 different controllers, which was done before." But the control engineer must consider practical matters like cost and development time. "In many cases it is easier and faster to realize the controller in a rule-based fuzzy language."

Siemens is now working on fuzzy subways as well as fuzzy traffic control systems and cement kilns. "Another area we're interested in involves the banking business: recognizing signatures to a high degree," says Reinfrank.

The firm is building a fuzzy task force, which had six people at this writing and will eventually exceed ten. "That's the kernel of the group. We plan to train people in fuzzy logic and send them out again," says Reinfrank. "The first products we're aiming at are two to three years ahead."

Siemens is not alone. In 1991, the Italian firm SGS-Thompson Microelectronics launched a five-year $30 million program to develop fuzzy hardware. The company is a leading maker of microcontrollers for other products, and Gianguido Rizzotto, director of advanced systems architecture at SGS-Thompson, estimated that semiconductor sales in fuzzy logic would reach $1 billion by 1995.[1]

Peugeot has commenced a major project on the Car of the Future, and as part of it, Elie Sanchez is developing a fuzzy braking system that stops automatically. Before every stop sign, transmitters in the road will relay signals about distance and slope. The fuzzy system will size up this information and come to a smooth halt. Peugeot is now simulating this device.

Next, Sanchez and his group will tackle vehicle avoidance. "You follow some vehicle and if it stops suddenly, you can avoid the collision." Just as airplanes can land automatically even in dense fog, he says, cars will perform similar feats. They could also have screens with infrared cameras, offering a sharper view than mirrors. "There are many applications in the car," he says.

In China, Wang Peizhuang went on to develop set-value statistics, a means of rendering fuzzy sets with real-world data. Whereas probability expresses data as points, he says, set-value statistics uses ranges—a line on a chart instead of a dot. "The point is very precise and the line is flexible," he says. "Most of the time we can't get data as precise as a point. For example, what's the temperature about now? It's between 26 and 28°C. So set value statistics represents the real world very flexibly."

He has also recently developed a concept he calls *factor space theory*. "The selection of the universe of discourse is more important than the membership functions," he says. For example, to express *young*, one employs not just one universe of discourse, but many, such as face, personality, activity level. "A lot of factors determine *young*," he says. Though his set-value statistics has become well-known in China, he believes this approach will prove even more significant.

Though China has by far the largest number of fuzzy workers in the world, it has no fuzzy firms. Instead, the government assigns engineers to carry out projects and they use fuzzy logic when it helps. For instance, it is now controlling furnace temperature in steel mills and helping manufacture the plastic film that farmers place on the ground to prevent freezing. Programmers have also developed expert systems in Chinese medicine and weather forecasting, and Wang calls the latter "highly accurate." Weather forecasting is especially important in China because it remains 80 percent agricultural. And expert systems in large institutes prophesy where the climate and soil is best for planting oranges or other crops.

Yet the anticommercial bias of the nation has hidden many of these achievements from Western view. "I think in the future when China improves its market system, it will become the most developed country in fuzzy logic," Wang says.

AMERICAN LIFTOFF

In the United States, fuzzy engineering began with NASA. The space agency had funded Zadeh as early as 1966, though he says he received that grant because of his reputation in system analysis. "Had I said it was to work in fuzzy sets, perhaps I would have been turned down." It was the start of a significant collaboration. By 1983, NASA was giving him grants earmarked for fuzzy logic.

In 1985, Bob Lea, an aerospace engineer at NASA's Johnson Space

Center, was investigating ways to control vehicles on the surfaces of the moon and Mars. These terrains are often rugged, yet radio signals from earth take so long to reach the moon that terrestrial control is not feasible. The rover must somehow proceed over rocks and rifts without help from Houston. Lea knew nothing of fuzzy logic, but after reading a few articles on it, he thought it might help.

Like Østergaard, he found to his surprise that higher-ups fretted about image. "For my first year, my supervisors were saying, 'When you try to sell this, don't call it "fuzzy logic." You ought to come up with another name.' So we experimented with other things, like 'nondeterministic reasoning.' They thought that sounded better. At any rate, I went back and said I'll call it 'fuzzy reasoning' because that's what everyone calls it."

Since then he has worked on several different fuzzy systems to guide vehicles across the lunar and Martian surfaces. He has also created a simulation of a fuzzy docking system that could serve both the space shuttle and the space station *Freedom*. Instead of crew members handling these linkups, fuzzy logic would manage them automatically.

In addition, he has used fuzzy logic to smooth out problems that arise with tethers. Astronauts often want to send vehicles out into nearby space, and they link them with a line to keep them from drifting into the void. But the tether itself may oscillate or otherwise act erratically when the engineering and sensor data aren't perfect. Fuzzy logic straightens it out.

Moreover, Lea has used fuzzy logic for the shuttle's star-tracking system, to prevent it from locking on to false targets like ice debris. A camera tracking system based on fuzzy logic, another project of his, will ease traffic management around the space station. As the United States moves toward more unmanned space flights in the future, fuzzy logic could also help control a shuttle by replicating and learning from pilots' behavior.

Through a few researchers like Lea, NASA became a pioneer in fuzzy logic. But private firms for the most part clung to the coast. Masaki Togai, having seen their caution first hand, decided to steer his own ship.

THE FIRST FUZZY COMPANIES

After Togai failed to interest Rockwell in his fuzzy chip, a fellow researcher suggested he take a week's vacation and seek investors for a company of his own. He took this advice and, in late March of 1987, flew to Tokyo. He felt Americans believed that fuzzy logic amounted to "fake science," and he had to find backing elsewhere.

There he met with a friend at Core Corporation, who had been seven years his senior at the Defense Academy. They went to a tiny, elite sushi bar in an alley off the Ginza and talked for three hours. Togai outlined his ideas and said that many firms wanted the hardware to implement fuzzy logic because of the current limits on computational power. In fact, his friend at Core needed Togai's help too, since he was creating an R&D group at the firm.

As a result of this conversation, Core promised Togai $150,000 in seed money for the first round of funding. He was launched. With that impetus, he began contacting other organizations.

A prime target was Canon. It took almost a year to see Canon executives. Eventually, a friend of his talked to a member of the board of Canon. "He saw that fuzzy logic was important for the future of Canon, and they approached me," Togai says. By then, company president Kazo Yamaji realized that fuzzy logic would be a key technology for autofocus cameras and the firm had begun work in this area. Canon welcomed Togai and infused financial life into his enterprise.

In his first year, he secured about $1.2 million. "Core was easy," he says, "but until Canon came in, it was difficult. We visited a lot of places." Canon now owns 18 percent of his firm. It also gave him numerous contract jobs, involving tasks such as autofocus and chip design.

In 1987 he founded Togai InfraLogic in Irvine, California, near the semiconductor product division of Rockwell in Newport Beach. He could have stayed in Japan and worked there, but opted to build an American company. "We can get better engineers in the United States," he says.

At first, his customers were solely Japanese. However, he has become involved with an increasing number of American firms, including auto companies and elevator firms. He is also working closely with NASA, which is using his fuzzy development system. Though not directly involved with the Mars Rover, he is providing tools and hardware for it. Togai also sells a fuzzy expert shell and add-on cards for regular computers.

Togai's fuzzy chips could find their way into many areas of everyday life. For instance, he says they could cut the wait in checkout lines while distant computers scrutinize credit cards. If the computer has enough information, he says, it can usually report back quickly. If it does not, a human operator must step in, causing delay. "In most cases, that wait cannot be longer than 15 seconds or people will get irritated," he says. This problem looms large in busy seasons like Christmas, when shopping lines lengthen and demands on the database intensify. A fuzzy chip, he says, can make a reasonable guess based on partial information, slashing time and soothing holiday nerves.

Zadeh foresees a central role for fuzzy chips in expert systems. "In most cases of control, the rules are simple," he says. "But as soon as you move into the domain of expert systems, diagnostic systems, nuclear reactor control, anywhere exceptions may play an important role, then simple software no longer suffices. So we may move to fuzzy logic chips, particularly since the cost is coming down very rapidly. So it was $200 or $300 and now it's $5 and soon it'll be $1 or $2." Hitachi is working on such special-purpose chips, and Zadeh predicts they will appear by 1993 or 1994.

As interest in fuzzy logic rose in the United States, Togai InfraLogic quickly gained recognition as the preeminent fuzzy firm in the United States. In 1989 *Electronics Products* magazine named its fuzzy processor "Product of the Year," and Togai opened branches in Tokyo and Munich. Even so, in the United States he was almost alone. Almost, but not quite.

A few other enterprising engineers soon started fuzzy firms as well. One, Aptronix, began in China and now conducts its affairs on the sunny flats of Silicon Valley.

Its founder is Wei Xu ("way shu"), a tall, slender man who left China in 1986 with degrees in engineering and business, and appears quite Westernized in the elegant suits he likes to wear. Wei had studied with Wang Peizhuang, who says, "The system in China is not so good. If the country changes, there'll be more people like Wei Xu." In the 1980s, Wei worked as a bureaucrat in the State Planning Commission, supervising control systems in power plant construction. "I always wanted to form a company in China doing fuzzy sets, but we didn't have the opportunity and we didn't have the funds," he says. "You know we are not free."

Yet by 1986 China was changing fairly rapidly. "In Japan, fuzzy was very popular, fashionable," Wei says. "I thought maybe it would be a good place to start up." He went to Japan and worked as a consultant, shuttling back and forth frequently from China. All the while he was ramping up to start his own firm. "If you want to have a company, you have to have something to sell," he says. "I was trying to figure out strategy."

In 1988, he launched Aptronix with $80,000 in borrowed funds. He planned to design industrial controllers and within a week had signed a contract worth over $100,000 with a major Japanese firm, receiving half in advance. "I knew we had good technology," he says. That deal gave him time to develop new products and he and his small staff went to work building a variety of fuzzy controllers for such applications as running a production line and controlling highway repair machinery.

Though his business prospered, Wei faced problems mastering Japanese and maintaining his visa. He respected the American engineers he met in China and Japan, and admired the U.S. computer industry. So he simply moved to the United States—without ever having seen it and hoping to build a career on pariah technology. At the outset, like Togai, he found most of his clients in Japan, but he was soon aggressively pursuing deals with companies like General Motors, United Technologies, McDonnell Douglas, and Texaco.

Engineer Xiwen Ma joined him in the United States in 1989, after the bloodshed at Tiananmen Square. He had been attending an AI conference here and decided to stay. He has since tried to bring his wife and two children over, but says strict U.S. immigration laws block their entry. "So if I cannot move my family here by 1993," he says, "I have to go back."

In the late 1980s the market for the powerful desktop computers called *workstations* was booming. These machines ran as fast as the mainframes of a decade earlier and were unexcelled at graphics. Wei jumped into this field with a fuzzy inference board that quickened calculation on a Sun workstation, and a fuzzy 3-D graphics adapter that he says increased speed 10 to 15 times.

Aptronix also developed the inverted triple pendulum, using fuzzy logic to balance three poles atop each other. "The triple pendulum is a problem that modern control cannot solve," Wei says. "It's impossible to solve all the differential equations in real time. When we use fuzzy, we can represent the problem in the rules, rather than in the equations, so we can solve it." So far it works only on a computer screen, in simulation. Wei says fuzzy logic could just as easily control three genuine poles, but it would be costly, requiring a large model with very rapid reflexes. Nonetheless, even in simulation, no other control techniques have matched it.

Wei envisions another first: a computer language for hardware designers who lack formal programming skills. It could let them create software for items like TVs and telephones, and thus realize their ideas directly. "It should result in easier-to-use VCRs," he notes. He foresees fuzzy logic improving cellular phones, control equipment, automobiles, data compression, and image processing in computer graphics and multimedia. "There's a huge market in the U.S.," he says.

Fred Watkins is a tall, bearded man who became interested in neural nets in 1987, when they were just emerging from the oubliette and promised an exciting future. He founded a company that sold neural net products, but soon saw a void under the glow. "It was my opinion they had no immediate use," he says, "because they don't do what people want to use them for."

He took a course from Bart Kosko at UC San Diego and discovered fuzzy logic. "Bart went on about fuzzy, about how wonderful it is," Watkins says. "I don't think it's the be-all-and-end-all, but it has a lot of pluses." He studied it and concluded that it was fast, maintainable, and, most important, practical. It yielded real solutions.

So he founded HyperLogic and began selling the PC program Cubicalc, which takes its name from Kosko's hypercube. Cubicalc is an environment that allows one to create and run a fuzzy expert system. "Let's say you're a financial wizard and you have an idea, albeit vague, about how the stock market works," he says. "Say you could summarize your thoughts in a few words. 'If the market opens high, then do this.' With Cubicalc you can put those sentences into the program and tell it what you mean by those nouns." Cubicalc could then fire the rules and reach a decision: Buy or sell some amount of some stock.

"Fuzzy logic is good for, say, parking cars, where people do an excellent job, but not to the millimeter," he says. "Where a good solution is obtainable, but not to super-high accuracy. A financial analyst can take some losses once in a while, as long as most of the time he shows a profit."

*   *   *

While working on his Ph.D. at UCLA, Kamran Parsaye was a research assistant for Judea Pearl. A witty, tactful man, Parsaye describes himself as "an innocent bystander" in the battle between fuzzy theorists and probabilists. "I try to avoid the intellectual gunfire," he says. But he ultimately could not ignore fuzzy logic. It was essential for his firm's flagship database.

Parsaye's company IntelligenceWare has developed a desktop computer program that combs large databases and finds patterns in them automatically. It turns a database from a passive repository of information into an active program that automatically asks questions of the data. It uses both Bayesian probability and fuzzy logic, and he says it "couldn't survive without either." Statistics form the core of the program and fuzzy logic helps perform discovery and communicate with the user.

The U.S. Forest Service is already using the software to predict the potential damage of a fire to a forest. It analyzes how flames would scar different areas in the woods and suggests preventive measures to lessen the harm. The program has also helped the USC Cancer Registry in Los Angeles correct the statistical methods it used to study lead poisoning. Surprisingly, the software revealed gender differences in lead toxicity, which people and other statistical programs had overlooked.

Though his software uses fuzzy logic, Parsaye does not broadcast the fact. In fact, the company's brochures don't mention fuzzy logic at all. "We really have no stake," he says. "We don't care what we're using internally as long as people are buying the program." But could he have created the same software with probability alone? "I don't even want to think about it," he says. "It's also possible to hunt a rabbit by hitting it with a gun. You'd have to run very fast."

THE WEIGHT AND THE TUG

Few large U.S. corporations paid much heed to fuzzy logic in the late 1980s. Ironically, the major exception was Rockwell, which had earlier brushed aside Togai's fuzzy chip. "We didn't make a mistake in not going as fast as Masaki wanted us to go," says Allen Firstenberg, director of information sciences at Rockwell's science center, who did not participate in this decision. "We needed to look more closely and I'd do that today. I personally would have liked us to commercialize certain products earlier than we have." But because of its early exposure, Rockwell is today one of the most advanced U.S. firms in the field.

It now markets several fuzzy products. Its Allen Bradley division sells a tuner that oversees the workings of a traditional controller. Rockwell is also testing a fuzzy logic flight control system to regulate active wing surfaces on planes. "You want to be able to roll the aircraft at high speed and fuzzy allows you to achieve that roll rate faster than any modern conventional control rate," says Firstenberg. With NASA, the company also is using fuzzy logic to regulate power management in the space station. A conventional system would assess the needs in different areas, such as a communications module and a life support module, and either send energy to them or not. A fuzzy system would be more flexible, perhaps running modules at various levels of power.

Outside Rockwell, however, U.S. corporations dawdled and yawned. "Engineers had heard about fuzzy logic and it's easier to ignore things. You hear about so many things," says Steve Marsh of Motorola. "The initial reaction is: 'I don't want to know about it.' Then when the information has been made available to you, you go through the skepticism phase: 'Yeah, but it has these drawbacks.' Then comes the phase when you see things go into real products. Now you say, 'Will I become obsolete by not knowing this?' That's the psychology of engineering Motorola has been involved with for some time."

The first-phase inertia was weighty. Classic control technology had served engineers well enough for a long time and it formed the foundation of their training. "I've found that a lot of Americans have a notion that if they've done something for 10 to 20 years and you change it, you've repudiated their whole life," says Doug Holmes, manager of advanced information technologies at Pacific Gas & Electric (PG&E), the California utility.

Kim Eckert, a manager of new products at Motorola, says, "Engineers are reluctant to look at a new approach, and even adamant about it. They're very vocal and protective of what they see as the only way of doing it. They feel threatened. They spent years studying and putting in practice the theory of well-defined and proven PID functions and controllers."

At a time when American competitiveness in high-tech faced erosion almost across the board, Zadeh found this laggardly response to Japan astonishing. "People are very good at rationalizing inaction," he said in 1991. "You see, it's part of a pattern which resists innovation. And American industry has unfortunately become afflicted with that. The people in the United States have lost a certain entrepreneurial spirit."

Sheridan Tatsuno notes that Americans show a reluctance to learn from others, just the opposite of the Japanese. "There's a whole set of mental rigidities and Americans have almost the worst attributes: Condescension to anyone not of their ilk," he says. "They shut off all these great ideas. You saw it in Egypt and Rome and China. It's a lack of openness and curiosity about the world. I see that more and more." Such complacency thrives at the core of prosperity, hollowing from within. In technology, where knowledge is wealth, it is a fast track to shantytown.

When past the stage of inert resistance, some engineers questioned the stability of fuzzy logic. Though it normally works fine, they worried that under exceptional circumstances it might go berserk. Japanese researchers say they test their products stringently, and the numerous trials of the Sendai subway and fuzzy appliances might seem to prove its dependability. However, U.S. engineers said that Japan had weaker product liability laws, so fuzzy mishaps might have occurred that the West had simply not heard about.

Sujeet Chand is manager of the control and signal processing department at Rockwell International. "I don't think there's a real problem with stability," he says, "but people in the West, especially in the United States, are so analytical that if you don't have a good handle on the stability of a controller and robustness, then they're unwilling to use it." In fact, there are intrinsic difficulties in determining whether nonlinear systems are stable. Chand has sought to resolve the problem by analyzing each mapping between input and output. If he can prove that each one is stable, he says, then the collection is stable.

Even engineers intrigued by fuzzy logic faced obstacles. How did it work? One could walk into a good Japanese bookstore and find six or seven introductory books on fuzzy logic right on the shelves. As late as mid-1992, there were none in the United States. Communication is vital to change paradigms and devise products, and the rising interest highlighted the near-void of information.

Marlin Eller of Microsoft stresses the need for practical papers, tutorials that engineers could follow. "There is a set of techniques that come out of fuzzy logic that can be very useful," he says. "But it's been difficult to find papers that give a practical feel for how to do things. I don't feel there are very good references."

Such works were all the more important since universities offered so little instruction on fuzzy logic. Ken Curry, a computer scientist at Eastman Kodak who became interested in fuzzy logic in 1990, complains that no one even mentioned it in his course work in AI. Few computer science or electrical engineering students were graduating with any exposure to fuzzy logic.

Some background in fuzzy logic would allow researchers to employ its concepts even if they eschew its mechanisms. Eller is development manager for Microsoft's Pen Windows, a software system to control pen-based computers and recognize handwriting. He did study fuzzy logic in his university days. "So I feel that in some of the ways we are working on the project, it's affected the way I think about things," he says. "We've not formally done anything that is fuzzy logic, but informally we have followed some of the basic tenets." Eller, however, is an exception.

The absence of engineering information led to an odd phenomenon: prods to innovation from journalism. Starting in the late 1980s, numerous U.S. magazines and newspapers ran stories on the boom in Japanese fuzzy products. "With so many articles in the popular press, the high executive will have seen something in *Business Week* or *Road and Track*," says Lee Feldkamp, a staff scientist with Ford. "He'll ask some person reporting to him, 'What the devil is this stuff?' This then trickles down."

In the United States, such top-down spurs to new technology are unusual. Generally, advanced technology groups seek out new ideas and inform management of the best techniques. Because rumor had tarred fuzzy logic and solid information was scarce in English, the usual path did not open up.

The main obstacle lay in the middle. "The control community has a great reluctance to accept this and, at least at Ford, that's the community that dominates the decisions in terms of control," Feldkamp says. "There's a lot of interest in the grass roots public in fuzzy logic. People were asking me about it before I knew anything about it.

Somewhere it has to meet in the middle where decisions get made, and at the moment that's where the problem is."

Even so, by the summer of 1991, the Japanese achievements had clearly piqued American interest. For instance, NCR was examining its utility in handwriting recognition and controlling paper movement. "If fuzzy logic is deemed to be appropriate, then there is no issue about going ahead," said NCR engineer John Vieth. "There isn't any reluctance."

Even so, Zadeh said at the time, "It will be extremely difficult for us to catch up. Japan is so far ahead," And some companies had become quite concerned. Neena Buck, a consultant then investigating fuzzy logic for several major corporations, said, "My clients are worried."

THE KNEE IN THE CURVE

No single event altered consensus in the United States like the second IFSA conference did in Japan. Rather, engineers and corporations changed their minds slowly and by degrees—a classic fuzzy transition. Yet if any one occasion marks the turning point, it took place in late June of 1991, in the sweltering humidity of Austin, Texas. The Microelectronics and Computer Technology Corporation (MCC) staged a two-day conference on fuzzy logic and over 300 people attended—engineers and executives from major computer, semiconductor, auto, oil, and many other firms. Indeed, MCC had to turn applicants away. It was the first time a crowd of that size, representing so many key American firms, had gathered to learn about fuzzy logic.

MCC is not an ordinary company, but a research consortium, hence a rare creature on American soil. It arose in 1983 to meet the Fifth Generation challenge and was, in fact, the first major U.S. joint venture in high-tech research explicitly based on Japanese models. Twenty firms—including 3M, Sperry, Rockwell, RCA, NCR, National Semiconductor, Motorola, DEC, Lockheed, and Eastman Kodak—joined as charter members and contributed capital, scientists, engineers, and directors. Among its original projects were a ten-year program focusing on parallel processing and AI, including neural networks.

Most U.S. firms remain wary of cooperative research, so such consortiums have tended to straggle. And trouble beset MCC early. A few key firms withdrew or merged, and the Fifth Generation went from monster to harmless phantom, draining urgency from the project. But despite occasional complaints that it had not delivered major break-throughs, MCC has forged on. With an annual budget of $60 million and the recent addition of Sun Microsystems, Apple, and Tandem Computers, it may yet play a key role in U.S. high-tech.

MCC first began looking at fuzzy logic in Japan in the late 1980s. "We were knocked off our feet by the applications and the speed of development and ease of maintenance of the applications," says MCC's Steve O'Hara. "You give this technology to an engineer and before you know it they've got a system functioning."

Yet Americans balked. "We spent some time just trying to convince people here and we ran into some hard times," he says. "We had a bit of an education process." To clamp the jumper cables on American interest, he and a few colleagues decided to hold a fuzzy logic conference.

The surprising turnout showed the new curiosity about fuzzy logic in the United States, a curiosity whetted partly by Japan and partly by the difficulty of finding clear and useful information on this subject. Bart Kosko, Takeshi Yamakawa, Tomohiro Takagi, Masaki Togai, and many others gave presentations, and Lotfi Zadeh regaled listeners with a colorful after-dinner speech. At any point, foes of fuzzy logic had the chance to grill these experts and perhaps embarrass them before a large and influential crowd, yet no challenge emerged. Indeed, when Ya-makawa showed a film in which he balanced a pole with a small platform at its tip, placed a wineglass atop it, and filled the glass with wine, the audience broke into applause.[2] (Yamakawa said later that he gave this particular demonstration "because my mind was occupied by a new wine that would arrive on the market in Japan. So I had an inspiration.")

Fuzzy logic impressed many in the audience. Doug Holmes of PG&E had been working on a project to automate the distribution of electricity. He had happened to read an article on Zadeh. "My boss and I were laughing, saying it would be incredibly useful, because you don't need to know everything. In the case of electrical distribution, you don't always have all the information." He attended the MCC conference, came away even more excited, and contacted Zadeh. "We'll definitely be doing something along that line," he said.

Others remained tentative. "We're trying to figure out where there might be an application, and I think in the United States there's some question of where the application is," said Ernie Green, a chemical engineer in the advanced technology group at Texaco. "I think we need to watch it and see what's going on. We're in the show-me stage. We haven't written it off. It's intriguing."

Yet many attendees were already convinced.

ON THE UPRAMP

The greatest presence at the MCC conference was Motorola, which sent some 50 engineers. Motorola had gone beyond the perusal stage and was actively educating its people, and its commitment stemmed not just from enthusiasm for the new technology, but from an unusual kind of market research.

In 1991 Steve Marsh, a director of marketing at Motorola, became associated with the firm's Center for Emerging Computer Technologies. Engineers there had been working on two neural net chips, code-named Batman and Android. "I saw what was occurring," he recalls, "and I said, 'Batman is great, but it's out of sequence.' It had some stuff good for the future. But I still had a responsibility to create a marketable chip and make lots of money."

He realized that fuzzy logic was accelerating faster than neural nets. "Customers wanted it. Being a marketing person, I believe if you have to work hard to market something, you should be marketing something else." Buyers of Motorola's standard microcontrollers, he knew, were starting to add fuzzy onto the chips.

How did he know this? In the past, microcontroller vendors have often not understood how buyers planned to use their chips. "That's one of the things that has plagued the manufacturers," he says. "It makes it harder to understand what they'll want in the future." Partly as a result, Motorola commenced a project to custom-design chips to buyers' needs, gratis. "We went to our big customers and told them, 'We want to do the design and we won't charge you and we'll give it to you in time and it'll cost less.'" This program put Motorola in direct touch with customer demand, "and they told us about their use of fuzzy logic and requirements for the future." Motorola managed to tap into this vein because many of its customers were Japanese.

Without such demand, Motorola would probably not have plunged into fuzzy chips, even if it recognized their value. "You can't introduce anything too soon because even if it's viable, it will seem to be a fad if people won't accept it," Marsh says. "If you become a fad player, you become a niche player. We don't want a niche. We want the whole thing."

Motorola personnel dragged their feet at first, but the tide from Japan soon caused consternation and attitudes began to change. "As fuzzy logic took off," Marsh says, "we accelerated our fuzzy logic activity and started associations with people very fuzzy literate. We had to bootstrap our effort quickly."

Around November of 1991 Motorola created an educational kit of its own, including a video. "It was hard to read about fuzzy logic in the texts available, written by mathematicians for mathematicians," Marsh says. "I had to go back to high school texts to find out set theory symbols. We thought it important to create an education course written by engineers who had applied fuzzy systems to give them rules of thumb."

Around the same time, programmer Jim Sibigtroth wrote a tiny nugget of software, 300 bytes in all, that took crisp input, looked for membership functions and rules stored as data, and performed fuzzy inference. "It worked well," says Marsh. Sibigtroth put it on Motorola's freeware line, where customers could access it and download it for free. The calls started pouring in. "Overnight, people were in the fuzzy logic business," Marsh says. "It became the most popularly downloaded item."

Motorola did not junk Batman and Android, but expanded them. It is now working on a chip that can carry out the functions of both fuzzy logic and neural nets. "That was one of ties we saw between those technologies," Marsh says.

By the time of the MCC conference, fuzzy logic was already wending its way through several industrial divisions of General Electric (GE). Researchers were looking at installing it in power systems to start and stop steam turbines more efficiently, and at using it to control the motors of steel rolling mills and jet engines.

But GE too passed through a phase of doubt before it began working on fuzzy appliances. Well before the MCC conference, the world's largest electronics company had taken note of the Matsushita fuzzy washer. "It got so much press coverage in magazines and newspapers that we ended up with a lot of people saying, 'What is this?' " says Steve Rice, an electronics development engineer at GE.

As a result, engineer Piero Bonissone set up a "fuzzy day" at GE. The program consisted of a seminar at which Bonissone and Vivek Badami, a computer and systems engineer, introduced fuzzy logic and explained its virtues. Some 40 people appeared, a turnout they considered good. "One of the reasons people showed up was the spate of articles on the Japanese efforts," says Badami. "It was a receptive but skeptical audience. There was a good deal of excitement, but no one was sold completely." A typical comment was: "Why do I need fuzzy logic?"

After the seminar, meetings took place where Bonissone and Badami told lead engineers that GE was considering using fuzzy logic in appliances. Badami thinks it makes sense. He says the Japanese washers that make decisions on the type of clothes, amount of soap and water, and gentleness of cycle would require between 1,000 to 2,000 rules in a standard PID controller. The fuzzy machines achieve the same results with 200 rules. "The point really is to reduce the level of effort. So it's a productivity gain you get with fuzzy logic," he says. "If on the one hand you have a tool that lets you do it with 200 rules, it's silly to implement it with 1,000. It's just a nice programming and implementation paradigm that is not available with traditional logic."

At the MCC conference General Electric did not seem alarmed at the prospect of vying with Japanese fuzzy washers. "I think a lot of the claims in Japan are overrated," said Rice. "There is fuzzy in them, but the level of fuzziness is not as deep as you're led to believe."

General Electric also wondered if American consumers would accept one-button automation. Its market research showed that buyers were used to paying less for products with fewer options and more for products with many. The more buttons, the better, Americans seemed to think. "When they spend $800, people want to make it do what they want, even if it's wrong," Rice said.

Rice added that the Japanese could not simply ship washers over here because their washers worked differently. They were smaller and lacked a central spindle. Moreover, he noted that fuzzy washing machines in Japan cost more than American consumers were willing to spend and that washing machines were simply not worth shipping across the

Pacific. They were bigger than TVs, and their profit margin did not permit the extra expense.

He may be right, but he may also be underestimating both the Japanese and the American consumer. For instance, would American buyers really reject one-button intelligence? The idea is novel here, and U.S. advertisers, so adept at selling illusory boons, might well rise to the challenge of a real one. Is it really important how washers work in Japan? Matsushita can obviously design an American-style machine. Does it matter that washers cost more in Japan than the United States? Almost all Japanese products cost more in Japan. Akihabara is no Crazy Eddie's. Japanese consumers subsidize their firms and have for decades. The classic pattern is simple: Companies test-market a product in Japan, work out the bugs, and build a fat profit margin. Then they slash the price and sell it to the larger American market. Japanese prices do not forecast American ones.

Finally, does the cost of trans-Pacific shipment matter if Japanese consumers subsidize it? It may be irrelevant anyway, for the Japanese pursue market share. They have not hesitated to offer extraordinarily low prices, toxic to American companies with small bankrolls. Yet once they control the market, prices drift up again and they recoup their losses. Shipment costs may not matter if you know that soon all your competitors will be Japanese.

Despite such hesitancies, by the spring of 1992, GE had plunged into fuzzy appliances. "That's sort of our major focus for fuzzy logic," says Badami. "It's still in the works. We're looking at it for dishwashers, washing machines, and refrigerators. It's not a state secret. Everyone else is too."

Badami notes that appliances require little precision. "The temperature for refrigerators doesn't have to be accurate to 0.01 degrees," he says. "So the demands aren't too high and the scope for fuzzy logic is quite good. You don't need a lot of time looking for high-precision models where you don't care about precision."

Moreover, cost and sensor efficiency also work against precision. "In washing machines, the measure for how clean your clothes are isn't that clear," he says. "You're looking at a $1 sensor and it has inherent imprecision. Given that sensor information is inherently imprecise, why worry about precise models in the process?"

Sophisticated math models are overkill, he says, because the expertise is mainly human. "You know if you use this much soap and keep it on long enough, that you'll get this result." Moreover, since the control is based on words, "it makes it easier to understand what you're doing, so you can extend it later. Say you added a humidity sensor to a refrigerator with a temperature sensor. You might add a few rules to the controller, like: If humidity is high, defrost more often. If you had to go back and model what refrigerator humidity was mathematically, it would probably take more time."

When GE begins selling fuzzy appliances, it probably will not label them as such. "In the United States we can't call anything *fuzzy* yet. *Fuzzy* still has such a negative connotation," says Rice. "It would be marketed as *smart* or *intelligent*."

Badami does cite drawbacks to fuzzy. For instance, it requires more programming code. Even so, he says, "Fuzzy is easier to write and think about and design." Moreover, a firm must pay an initial cost in coming up to speed in the technology. "It's a different way of thinking and writing equations. Thinking of rules and membership functions. And you pay an up-front price in doing that stuff," he says, though he adds that firms recoup this cost later.

Despite the increased acceptance at GE, he says, "There's still skepticism here. It's primarily from the control community. They're used to rigorous proofs and algorithms. It's annoying for them to see the community justifying itself on experimentation and empirical results. That's still a hurdle to overcome."

Badami and others have fought an "uphill battle" to gain funding and have especially sought successes to display. "What will convince people is not to say fuzzy logic will change the way you do things and make problems go away. If you approach it as just another tool, you'll go much further than saying this is a magical solution, which it isn't. It's another tool. A useful tool, but just a useful tool."

Whirlpool also makes washers and refrigerators, and it too is facing the Japanese challenge. Gerald Eisenbrandt is a Whirlpool staff engineer who helps monitor new technology. He had heard of fuzzy logic in the late 1960s, but had not looked further into it until the late 1980s, when the Japanese products caught his eye.

Many Whirlpool engineers initially found fuzzy logic somewhat alien and hard to adjust to. They also questioned how much it contributed to items like vacuum cleaners. "But you'd look at other products, like that Matsushita fuzzy camcorder that takes vibrations," Eisenbrandt says. "There you can see an obvious problem and anyone who has used a home video camera knows it. You can see a benefit that has substance and is clearly visible." Such achievements, along with the overall avalanche of products from Japan, slowly changed the skepticism to conviction. "The appearance of products put something concrete to it," he says. It also applied a little pressure at Whirlpool's back.

Though Whirlpool has no fuzzy products available yet, Eisenbrandt notes they are in development. "I think in general, you'll see them coming out," he says. "Whether they'll be called *fuzzy* is a question for the marketing people."

The beleaguered U.S. auto industry also spotted fuzzy logic as a key field, and by 1991 engineers at Ford were requesting seminars and tutorials on it. "There's a lot of interest," said Lee Feldkamp at the time. "Whether that will translate into action is a damn good question." Skeptics at Ford raised the stability question. "That's what the control people hold up to keep the demons of fuzzy logic, people like me, away."

By the spring of 1992, however, Ford had begun active work "on two or three vehicle systems," said Feldkamp, guardedly. "To the point in a couple of cases of having fuzzy running in hardware on test vehicles. I think that is as far as I can go on that." But he did suggest that fuzzy logic was suitable for cruise and idle speed control, as well as power train control in general.

General Motors likewise has launched projects in fuzzy logic, and it too is mum about the details. In any case, Detroit remains far behind the all-purpose fuzzy system of Mitsubishi.

Otis Elevator also went through a stage of denial and doubt. When Mitsubishi introduced its fuzzy elevator dispatcher, some Otis engineers were skeptical. Otis already had sophisticated dispatch systems based on crisp logic, and certain individuals, says senior vice president Siddiq Sattar, felt the improvements claimed by the Japanese merely enhanced an inefficient system, rather than one like theirs. "Sometimes when the competition presents a new product or new feature, to make it look good, the improvements are not compared with the best system," he says. But in 1989 Otis decided to probe the technology anyway.

A typical high-rise building has a bank of six or eight elevators, and people make calls on it from many different floors. The system must quickly decide where to send each car. These decisions are important. "You don't want to be in a position where some poor passenger is hung out on a floor for a long time because the system is inefficient," says Sattar. "It's not just improving average performance, but also taking care of long-waiting passengers. There's nothing worse than the passenger who gets frustrated because an elevator doesn't come for several minutes."

Otis assembled a small team which examined areas where fuzzy logic might pay off. They found several. One lay in assessing the number of people inside the car. "When you're riding the elevator, the elevator weighs you," Sattar says. "And it has a good idea how many people are in the car."

Another was time of day, which dictates traffic pattern. "In the morning when people arrive at the office, there is an up peak. Most people are going upward. And there's interfloor traffic during the day, especially if the tenant has several floors and people are going up and down between them. Then there's the down peak that takes place at lunch or in the evening." The bounds of these up and down peaks are fuzzy, and the elevator must know when it is heading toward one or the other.

The number of people waiting at the call is also a fuzzy factor, as is performance estimation, that is, the extent to which the system focuses on reducing the longest waiting time, say, as opposed to minimizing the average waiting time. "We can tailor the system to account for all that," Sattar says.

When Otis completed this analysis, he says, "We knew there was opportunity." Yet resistance remained. "It wasn't emotional resistance, but intellectual," Sattar says. "People felt you could do whatever fuzzy could do with crisp logic, and that you should do it with crisp. That may still be true, but fuzzy logic may be a way of achieving the same objective more efficiently."

Like Matsushita, Otis addressed these qualms carefully. "You have to build a consensus and get people going in that direction, as opposed to slam-dunking a project," Sattar says. "You have to have the right people making convincing arguments. We probably talked about it for nine months to a year." But in late 1989 Otis went ahead, creating some modules which it superimposed on the existing system, and by spring of 1992 it was writing the software. "It'll be available to our customers in Japan in October of 1992," he says.

Japan? Not the United States?

Of course. Otis's buyers in Japan know of fuzzy logic and have asked for it. "The Japanese customer base is very technology-sensitive," says Sattar. "They're willing to try new things. It's a good place to introduce and expand from there into Europe and North America." In fact, Otis unveiled another new elevator product there in 1990, a kind of motor. "We see in Japan a customer base hungry for new products."

Moreover, unlike most U.S. companies, Otis has maintained a presence in Japan since before World War II, and has strong ties with Matsushita and the huge Sumitomo *keiretsu*. Indeed, Otis builds around the globe, and of its 50,000 employees, only 8,000 work in the United States. "If it weren't for our company in Japan," he says, "we wouldn't be as far along in fuzzy logic as we are."

YESTERDAY, TODAY, AND TOMORROW

Lotfi Zadeh's life has changed radically in the last few years. The rise of fuzzy logic has swept him up from the classrooms of Berkeley and placed his face in newspapers and magazines all over the world. He has become a well-known name in Japan, and in Europe strangers have stopped him in the street. Yet he remains unflappable. "I feel no different today than when fuzzy logic was in the doghouse and people said it was nonsense," he says. "Of course, I've become a different figure. I notice people treat me differently. But so far as I'm concerned, I don't feel any different."

He does, however, feel different about the prospects for fuzzy in the United States. "The situation is changing rather rapidly now," Zadeh said in mid-1992. "It has become a qualitatively different phenomenon."

Zadeh believes the United States at last has a chance to catch Japan. "I think the gap will be narrowing," he says. "But at this point the Japanese have something we don't have. They have a fairly large number of engineers in industry who have quite a bit of design experience by now. These engineers can be turned loose on all kinds of products." For instance, he notes, if Mitsubishi wants to infuse a car with fuzzy logic, it just issues the command, and before long a fuzzy car materializes. "We can't do that. Our engineers are in the training mode. So if you look at number of applications, the gap will be widening because they have those engineers in place. On the other hand, the thing will pick up steam in the United States, so in a year or so there will be people with design experience."

At the same time, the crippling information void is slowly ending. "In the near future we'll see a profusion of books on fuzzy logic, like introductory texts," says Zadeh. "I know of several now that are being written or reviewed."

Information can lead to rapid progress. Even a single university course can yield a trove of products. For instance, in the spring of 1992, students in Bart Kosko's class on fuzzy systems designed such feasible fuzzy devices as

- Insulin pumps.
- Incubators for premature infants, to help regulate their temperature.
- Devices to induce labor by smoothing the flow of anesthetic and the drug pitocin to the expectant mother.
- Pool chlorinators.
- Throttle controllers for racing powerboats, to keep their bows from pitching out of the water.
- Aquarium management systems.
- Heart pacemakers.
- Light dimmers.
- Liquid cooling systems for computer workstations.

Moreover, they had modeled more difficult tasks, such as landing autopilots for small aircraft and robot control. And these were students, not highly paid specialists in expensive labs.

Fuzzy logic requires no massive outpouring of cash as does, say, high-definition TV. It is simple enough to learn and, as one engineer after another has discovered, fast and easy to effect. It thus seems unusually well-suited for American firms, especially those facing a shortage of capital and a decline in consumer confidence compared to their Japanese rivals. These virtues mean that the West is not shut out of the fuzzy market. It can become competitive fairly quickly.

It is an opportunity. Mistakes litter the past. The West scoffed at

Zadeh's ideas and ignored Mamdani and Holmblad. Both AT&T and Rockwell spurned Togai and Watanabe's fuzzy chip. The Los Angeles subway went ahead in ignorance of the slickest train in the world. Until the Japanese, few American companies had even heard of fuzzy systems, and those that had dismissed it.

Yet Western civilization has overcome biases inherited from Aristotle in the past, and without the economic goad. And fuzzy logic is practical in the highest sense: direct, inexpensive, bountiful. It forsakes not precision, but pointless precision. It abandons an either/or hairline that never existed and brightens technology at the cost of a tiny blur. It is neither a dream like AI nor a dead end, a little trick for washers and cameras. It is here today, and no matter what the brand name on the label, it will probably be here tomorrow.

ACKNOWLEDGMENTS

■

We could scarcely have written this book without the cooperation of the scores of people we interviewed, sometimes under curious circumstances. For instance, we first talked to Takeshi Yamakawa in a cafeteria room at midnight and to Anca Ralescu in a restaurant beneath a Tokyo subway station, as waitresses bustled by. Every one of them, even the detractors of fuzzy logic, graciously lent us their time and expertise, and we appreciate it.

We owe special gratitude to Lotfi Zadeh, the origin and heart of fuzzy logic, and Bart Kosko, its philosopher and seer. They discussed the topic with us over and over again, so often that they ultimately transcended the role of interviewee. Without their patience and understanding, this book would be much poorer.

Our research in Japan posed the inevitable cross-cultural problems, and many people came to our aid here as well. Mana Nishimura lived up to her wizardly repute by obtaining an open sesame and setting schedules for us everywhere. Our debt to her is huge. The engineers and managers of the Japanese corporations were unfailingly warm and generous, as were the translators: Kiyomi Morishita of Mitsubishi Electric, Kazuaki Urasaki of Omron, Akira Nagano of Matsushita, Takaaki Iwata of NOK, and Riichiro Uyeda of Hitachi. And Dr. Yamakawa and his aides at FLSI—Hiroko Noguchi, Toyoko Kuniyasu, Eiichi Goto, and Tsutomu Miti—helped us arrange accommodations and sidestep myriad little difficulties. To all: *Iroiro osewa ni narimashita*.

A number of others cheerfully took time to advance our project. Beth Hughes, Yogan Dalal, Darren Robbins, Harry Pachon, Jerry Kaplan, Mike and Sue Elwell, Rob Polevoi, John Markoff, Howard Bailen, Jan Woleński, Piero Metelli, Sadayasu Shibata of Toshiba, and Brent

Schnell of Omron all came forth with interesting information. Fay Zadeh combed the records of her husband's active life and gave us context for many of its highlights. And Fred Trapkin, Shirley Yawitz, and Vic Bruni showed pluck and vigor in removing extraneous obstacles to this book.

We are indebted to Keay Davidson, Federico Faggin, Héctor Feliciano, Rosalind Gold, David Liddle, George Morrow, and Phil Robinson, who read all or parts of this manuscript and pointed out inaccuracies, longueurs, and obscurities. Their insights adorn the book. It goes without saying—but must be said—that any errors are our responsibility alone.

We also owe much to our agents, John Brockman and Katinka Matson, who were present at the inception; Joe Shoulak, who created many of the illustrations; David Leon and Paul Saffo, who came through in times of crisis; and our editor, Dominick Anfuso, who gently helped us streamline and sharpen the narrative.

Finally, our wives, Rosalind Gold and Jeanne Loveland-Freiberger, lived through the writing of this book, and they are the foundation on which it rests.

NOTES

■

Fuzzy logic is an international story, and in researching it we talked to scholars, engineers, and managers from all over the world. They include

Vivek Badami
Hans-Walter Bandemer
Hamid Berenji
Hans Berliner
Bob Brown
Neena Buck
Sujeet Chand
Peter Cheeseman
Ken Curry
John Dockery
Shingi Domen
Kim Eckert
Gerald Eisenbrandt
Doug Elias
Marlin Eller
Federico Faggin
Lee Feldkamp
Allen Firstenberg
Itsuko Fujimori
Yoshihiro Fujiwara
Toyō Fukuda
Brian Gaines
Ernest Green
Tadakuni Hakata
John Havens
Isao Hayashi
Kevin Helliker
Masahiko Hishida
Peter Holmblad
Doug Holmes
Alan Huang
Yasuteru Ichida
Victoria Kader
William Kahan

Kitahiro Kaneda
Yuzo Kato
Tadashi Katsuno
Paul Kay
George Klir
Bart Kosko
Bob Lea
Ramón López de
    Mántaras
Akira Maeda
Tadasu Maeda
Ebrahim Mamdani
Steve Marsh
Dominic Massaro
Jiro Masuda
Marvin Minsky
Shoji Miyamoto
Atsushi Morita
Gregg Oden
Michael O'Hagan
Steve O'Hara
Jens-Jorgen Østergaard
Masayuki Oyagi
Kamran Parsaye
Judea Pearl
François Pin
Henri Prade
Anca Ralescu
Michael Reinfrank
Werner Remmele
Steve Rice
Stephen Rodabaugh
Eleanor Rosch
Elie Sanchez

Siddiq Sattar
Takeshi Sawada
Yasuhide Seno
Michio Sugeno
Rod Taber
Tomohiro Takagi
Sheridan Tatsuno
Toshiro Terano
Masaki Togai
Richard Tong
Bob Townley
Jerry Vaughan
John Vieth
Noboru Wakami
Wang Peizhuang
Fred Watkins
Wei Xu
Andrew Whiter
Frank Wu
Xiwen Ma
Ron Yager
Takeshi Yamakawa
Chizuko Yasunobu
K. K. Yawata
John Yen
Terukuni Yokoyama
Tadao Yoshikawa
Tatsuaki Yukimasa
Fay Zadeh
Lotfi Zadeh
Maria Zemankova
Hans Zimmermann

Unless cited with a note, all quotations stem from these interviews.

287

PROLOGUE

1. Communication to Lotfi Zadeh.
2. In fact, the term *faaji*, otherwise meaningless in Japanese (*aimai* means "fuzzy"), has become synonymous with *clever*.

CHAPTER 1

1. Lotfi Zadeh, "Biological Application of the Theory of Fuzzy Sets and Systems," in *The Proceedings of an International Symposium on Biocybernetics of the Central Nervous System*, Boston, Little, Brown, 1969, pp. 199–206, p. 199.
2. Berkeley mathematician Hans Bremermann demonstrated this fact in 1962. Relying on quantum theory, he first determined how much information the most efficient computer imaginable could process each second. He found that, for each gram of mass, no data processing system—brain, machine, anything—could handle more than twice 10^{47} bits per second.

He then asked, "What if a computer were as big as the earth, and it had labored on a single problem since the earth was born? How many bits could it process?" His answer was 10^{93}, usually called Bremermann's limit. It is a huge figure—1,000,000,000,000,000,000,000,000,000,000,000,000,000,000,000,000, 000,000,000,000,000,000,000,000,000,000,000,000—but it still clamps a lid. In chess, for instance, there are some 10^{120} possible move sequences—a figure so much larger that even if each move sequence were a single bit, one could only compute 1/1,000,000,000,000,000,000,000,000,000,000th of their total number. See Hans J. Bremermann, "Optimization Through Evolution and Recombination," in M. C. Yovits et al., eds., *Self-Organizing Systems*, Washington, D.C., Spartan Books, 1962, pp. 93–106.
3. Warren Weaver, "Science and Complexity," *American Scientist*, 36, 1948, pp. 536–544.
4. Lotfi Zadeh, "From Circuit Theory to System Theory," *Proceedings of Institution of Radio Engineers*, 50, 1962, pp. 856–865, p. 857.
5. He also forced foreigners to call Persia "Iran." The ancient *Zend-Avesta* referred to the land as *Airya*, which related to *Arya* or "noble," from which Hitler derived *Aryan*.
6. In Milan Zelený, "Cybernetyka," *International Journal of General Systems*, 13, 1987, pp. 289–294, p. 290.
7. Lotfi Zadeh, "Thinking Machines: A New Field in Electrical Engineering," *Columbia Engineering Quarterly*, January 1950, pp. 12–13, 30–31.
8. David Marr, "A Theory for Cerebral Neocortex," *Proceedings of the Royal Society of London*, Ser. B, 176, 1970, p. 161.
9. In Morris Kline, *Mathematics: The Loss of Certainty*, New York, Oxford University Press, 1980, p. 201. For more on Cantor's life and work, see Joseph W. Dauben, *Georg Cantor: His Mathematics and Philosophy of the Infinite*, Cambridge, MA, Harvard University Press, 1979.
10. In Beverley Kent, *Charles S. Peirce: Logic and the Classification of the Sciences*, Montreal, McGill-Queen's University Press, 1987, p. 9.
11. In Nicholas Rescher, *Many-valued Logic*, New York, McGraw-Hill, 1969, p. 5.
12. In Mihai Nadin, "The Logic of Vagueness," in Eugene Freeman, ed., *The Relevance of Charles Peirce*, La Salle, IL, Monist Library of Philosophy, 1983, pp. 154–166, pp. 157–158.
13. In Brian Gaines, "Foundations of Fuzzy Reasoning," *International Journal of Man-Machine Studies*, 8, 1976, pp. 623–668, p. 660.
14. In Nadin, "Logic of Vagueness," p. 154.

15. Bertrand Russell, "Vagueness," *Australasian Journal of Psychology and Philosophy*, 1, 1923, pp. 84–92, p. 84.

16. Ibid., p. 89.

17. In Max Black, "Vagueness: An Exercise in Logical Analysis," *Philosophy of Science*, 4, 1937, pp. 427–455, p. 451.

18. Boleslaw Sobocinski, "In Memoriam Jan Łukasiewicz," *Philosophical Studies* (*Ireland*), 6, 1956, pp. 3–49.

19. C. S. Peirce, the Scotsman Hugh MacColl (1837–1909), and the Russian Nikolai A. Vasilev (1980–1940) developed earlier multivalued logics, but theirs lacked his impact.

20. Jan Łukasiewicz, "In Defense of Logistic," in L. Borkowski, ed., *Selected Works*, London, North-Holland, 1970, p. 246.

21. Max Black, "Vagueness: An Exercise in Logical Analysis," *Philosophy of Science*, 4, 1937, pp. 427–455, p. 451.

22. Lotfi Zadeh, "Fuzzy Sets," *Information and Control*, 8, (3), June 1965, pp. 338–353.

23. Lotfi Zadeh, "Making Computers Think Like People," *IEEE Spectrum*, August 1984, pp. 26–27.

24. Wilfrid Hodges, *Logic*, New York, Penguin Books, 1977, p. 34.

25. Bart Kosko, *Neural Networks and Fuzzy Systems*, Englewood Cliffs, N.J., Prentice-Hall, 1991, p. 267.

26. Clifford Geertz, *Local Knowledge: Further Essays in Interpretive Anthropology*, New York, Basic Books, 1983, pp. 19–35. See also Kenneth J. Gergen, *The Saturated Self*, New York, Basic Books, 1991, pp. 111–117.

27. Clifford Geertz, *Local Knowledge*, p. 20.

28. A. Cornelius Benjamin, "Science and Vagueness," *Philosophy of Science*, 6, 1939, pp. 422–431.

29. Lotfi Zadeh, "Outline of a New Approach to the Analysis of Complex Systems and Decision Processes," *IEEE Transactions on Systems, Man, and Cybernetics*, SMC-3(1), January 1973, pp. 28–44, p. 28.

30. Ibid., p. 29.

CHAPTER 2

1. Max Black, correspondence to Lotfi Zadeh, June 21, 1967.

2. In Lotfi Zadeh, "The Birth and Evolution of Fuzzy Logic," speech presented on acceptance of the 1989 Honda Prize, pp. 2–3.

3. Ibid., p. 4.

4. Lotfi Zadeh, "Making Computers Think Like People," *IEEE Spectrum*, August 1984, p. 26.

5. Ernest Nagel and James R. Newman, *Gödel's Proof*, New York, New York University Press, 1958, p. 53, italics in original.

6. In Robert Payne, *Ancient Greece: The Triumph of a Culture*, New York, W. W. Norton, 1964, p. 399.

7. Charles Hartshorne and Paul Weiss, eds., *Collected Papers of Charles Sanders Peirce*, 5, Cambridge, MA, Harvard University Press, 1934, p. 223, §358.

8. Aristotle, *The Metaphysics*, trans. Hippocrates G. Apostle, Bloomington, Indiana University Press, 1966, p. 60.

9. Ibid., pp. 58–59.

10. Ibid., p. 70.

11. Gottfried Wilhelm Leibniz, "The Nature of Truth," in G. H. R. Parkinson, ed., *Gottfried Wilhelm Leibniz: Philosophical Writings*, London, G. M. Dent & Sons, 1973, p. 93.

12. In the preface to the second edition of Kant's *Critique of Pure Reason*, cited in Heinrich Scholz, *Concise History of Logic*, New York, Philosophical Library, 1961, p. 1.

13. Aristotle, *Metaphysics*, p. 64.

14. Aristotle, *The Nicomachean Ethics*, trans. J. E. C. Welldon, London, Macmillan, 1892, p. 3.

15. Ibid., p. 4.

16. Thomas Kuhn, *The Structure of Scientific Revolutions*, Chicago, University of Chicago Press, 1962.

17. In Morris Kline, *Mathematics in Western Culture*, New York, Oxford University Press, 1953, p. 397.

18. Ibid.

19. Willard V. N. Quine, *The Philosophy of Logic*, Cambridge, MA, Harvard University Press, 1970, p. 84.

20. Ibid., p. 85.

21. Willard V. N. Quine, *Quiddities: An Intermittently Philosophical Dictionary*, Cambridge, MA, Belknap Press, 1987, p. 56.

22. In Morris Kline, *Mathematics: The Loss of Certainty*, New York, Oxford University Press, 1980, p. 88.

23. Stephen O'Brien, "The Ancestry of the Giant Panda," *Scientific American*, 257, November 1987, pp. 102–107.

24. Friedrich Nietzsche, *The Gay Science*, trans. Walter Kaufmann, New York, Random House, 1974, §111, p. 171.

25. For this line of reasoning, see Plato, *The Republic*, trans., G. M. A. Grube, Indianapolis, IN, Hackett Publishing, 1974, Book V, p. 139.

26. Ernst Mayr, *The Growth of Biological Thought*, Cambridge, MA, Belknap Press, 1982, p. 38.

27. Charles Darwin, *On the Origin of Species*, London, Murray, 1859, p. 44.

28. In Mayr, *Biological Thought*, p. 267.

CHAPTER 3

1. Zadeh is an accomplished amateur photographer, especially excelling in portraiture. Alexander Kerensky used a Zadeh portrait without notice or permission on the cover of his autobiography, and a shot of a stunningly devious Richard Nixon adorns the Zadeh home. Zadeh has also photographed such figures as Harry Truman, Edgard Varèse, Claude Shannon, C. Northcote Parkinson, and Mstislav Rostropovich.

2. Ray Simpson, "The Specific Meanings of Certain Terms Indicating Differing Degrees of Frequency," *The Quarterly Journal of Speech*, 30, 1944, pp. 328–30; Milton D. Hakel, "How Often Is Often?" *American Psychologist*, 23, 1968, pp. 533–534.

3. Lotfi Zadeh, "Fuzzy Sets as a Basis for a Theory of Possibility," *Fuzzy Sets and Systems*, 1, 1978, pp. 3–28.

4. Lotfi Zadeh, "Analysis of Complex Systems and Decision Processes," p. 28.

5. Lotfi Zadeh, "Fuzzy Logic and Approximate Reasoning," *Synthese*, 30, 1975, pp. 407–428.

6. Thomas Kuhn, *The Structure of Scientific Revolutions*, Chicago, University of Chicago Press, 1962, p. 169.

7. Eugene Frankel, "Corpuscular Optics and the Wave Theory of Light: The Science and Politics of a Revolution in Physics," *Social Studies of Science*, 6, 1976, pp. 141–184.

8. Thomas Kuhn, *The Structure of Scientific Revolutions*, p. 151.

9. Correspondence of Stephen Rodabaugh to editor Jerry Vaughan, September 30, 1982. Rodabaugh said Vaughan wrote to a fuzzy topologist: "The members of the editorial board with whom I have consulted feel that papers in pure fuzzy topology are outside the scope of this journal."

10. Michael Smithson, *Ignorance and Uncertainty: Emerging Paradigms*, New York, Springer-Verlag, 1989, p. 304.

CHAPTER 4

1. John Locke, *An Essay Concerning Human Understanding*, 2, London, Bye and Law, 1805, p. 45.

2. Ibid., p. 43.

3. David Hume, *A Treatise of Human Nature*, New York, Penguin, 1969, p. 64.

4. Eleanor Rosch, "On the Internal Structure of Perceptual and Semantic Categories," in Timothy Moore, ed., *Cognitive Development and the Acquisition of Language*, New York, Academic Press, 1973, pp. 111–144.

5. Eleanor Rosch, "Principles of Categorization," in Eleanor Rosch and Barbara Lloyd, eds., *Cognition and Categorization*, Hillsdale, NJ, Lawrence Erlbaum Associates, 1978, pp. 27–48, p. 40.

6. Ludwig Wittgenstein, *Philosophical Investigations*, trans. G. E. M. Anscombe, New York, Macmillan, 1953, pp. 31–34. The pages after this famous section treat the example of sharp and blurry pictures broached by Bertrand Russell. Wittgenstein knew Russell well, and much of this discussion seems to begin where "Vagueness" leaves off.

7. Eleanor Rosch and Carolyn B. Mervis, "Family Resemblances: Studies in the Internal Structure of Categories," *Cognitive Psychology*, 7, 1975, pp. 573–605.

8. Eleanor Rosch, "Principles of Categorization," p. 28.

9. Willett Kempton, "Category Grading and Taxonomic Relations: A Mug Is a Sort of Cup," *American Ethnologist*, 5, 1978, pp. 44–65.

10. Ibid., p. 61.

11. Paul Kay and Chad McDaniel, "The Linguistic Significance of the Meanings of Basic Color Terms," *Language*, 54, 1978, pp. 610–646.

12. Ibid., p. 636.

13. Michael Smithson, "Fuzzy Set Theory and the Social Sciences: The Scope for Applications," *Fuzzy Sets and Systems*, 26, 1988, pp. 1–21, p. 12.

14. Ibid., p. 3.

15. Ibid., p. 2.

16. George Lakoff, *Women, Fire, and Dangerous Things: What Categories Reveal about the Mind*, Chicago, University of Chicago Press, 1987, p. 141.

17. In Robert Burchfield, *The English Language*, Oxford, UK, Oxford University Press, 1985, p. xiii.

18. Howard Gardner, *The Mind's New Science: A History of the Cognitive Revolution*, New York, Basic Books, 1985, p. 354.

CHAPTER 5

1. Lotfi Zadeh, "Fuzzy Sets and Systems," presented at the Symposium on System Theory, Polytechnic Institute of Brooklyn, April 20–22, 1965, pp. 29–37, p. 37.

2. Lotfi Zadeh, "Analysis of Complex Systems and Decision Processes."

3. Ebrahim Mamdani and Sedrak Assilian, "An Experiment in Linguistic Synthesis with a Fuzzy Logic Controller," *International Journal of Man-Machine Studies*, 7, 1975, pp. 1–13.

4. Peter Holmblad and Jens-Jorgen Østergaard, "'Control of a Cement Kiln by

Fuzzy Logic," in M. M. Gupta and Elie Sanchez, eds., *Fuzzy Information and Decision Processes*, Amsterdam, North-Holland, 1982, pp. 398–399.

5. Hans Berliner, "Computer Backgammon," *Scientific American*, 242 (6) June 1980, pp. 54–60, p. 60.

CHAPTER 6

1. T. G. Kalghatgi, *Jaina Logic*, New Delhi, Shri Raj Krishen Jain Charitable Trust, 1984, p. 16.

2. Suzuki Daisetz Teitarō, "Reason and Intuition in Buddhist Philosophy," in Charles A. Moore, ed., *The Japanese Mind: Essentials of Japanese Philosophy and Culture*, Honolulu, University of Hawaii Press, 1967, p. 74.

3. Ibid., p. 95.

4. Joseph Needham, *Science and Civilization in China*, 2, Cambridge, UK, Cambridge University Press, 1954, p. 35.

5. Lao Zi, *Tao Te Ching*, trans., Victor M. Hair, New York, Bantam Books, 1990, p. 60.

6. Ibid., p. 13.

7. Ibid., p. 7.

8. Wing-tsit Chan, ed. and trans., *A Source Book in Chinese Philosophy*, Princeton, NJ, Princeton University Press, 1963, p. 244.

9. Ibid., p. 233.

10. In Laurence C. Wu, *Fundamentals of Chinese Philosophy*, New York, University Press of America, 1986, p. 190.

11. In Wing-tsit Chan, *Chinese Philosophy*, p. 233.

12. Hideki Yukawa, "Modern Trend of Western Civilization and Cultural Peculiarities in Japan," in Moore, *Japanese Mind*, p. 56.

13. Charles A. Moore, "Editor's Supplement: The Enigmatic Japanese Mind," in ibid., p. 290.

14. In Frederik L. Shodt, *Inside the Robot Kingdom*, Tokyo, Kodansha International, 1988, p. 212.

15. Sheridan Tatsuno, *Created in Japan: From Imitators to World-Class Innovators*, New York, Harper & Row, 1990, p. 22.

16. In Hajime Nakamura, "Basic Features of the Legal, Political, and Economic Thought of Japan," in Moore, *Japanese Mind*, p. 159.

17. Ibid., p. 160.

18. Yukawa, "Modern Trend," p. 56.

19. Clyde Prestowitz, *Trading Places: How We Allowed Japan to Take the Lead*, New York, Basic Books, 1988, p. 33.

20. This extraordinary character, a popular playwright, went on to achievements in electricity, botany, economics, and art. He founded a mine, produced sugar, and bred sheep for wool and died after incarceration for beating a student of his to death.

21. Sonni Efron, "A Cultural Trade Imbalance," *Los Angeles Times*, May 21, 1992, p. 1.

CHAPTER 7

1. For a more detailed account, see George Klir and Tina Folger, *Fuzzy Sets, Uncertainty, and Information*, Englewood Cliffs, NJ, Prentice-Hall, 1988, p. 107 et seq.

2. See Michio Sugeno and M. Nishida, "Fuzzy Control of Model Car," *Fuzzy Sets and Systems*, 1985, pp. 103–113; and Michio Sugeno, T. Murofushi, T. Mori, T. Tatematsu, and J. Tanaka, "Fuzzy Algorithmic Control of a Model Car by Oral Instructions," *Fuzzy Sets and Systems*, 32, 1989, pp. 207–219.

3. Kazuyoshi Kamioka, *Japanese Business Pioneers*, Union City, CA, Heian International, 1988, p. 122.

4. Seiji Yasunobu and Shoji Miyamoto, "Automatic Train Operation System by Predictive Fuzzy Control," in Michio Sugeno, ed., *Industrial Applications of Fuzzy Control*, Amsterdam, North-Holland, 1985, pp. 1–18.

5. Kaoru Hirota, Yoshinori Arai, and Shiroh Hachisu, "Fuzzy Controlled Robot Arm Playing Two-Dimensional Ping-Pong Game," *Fuzzy Sets and Systems*, 32, 1989, pp. 149–159.

6. Takeshi Yamakawa, "Stabilization of an Inverted Pendulum by a High-Speed Fuzzy Logic Controller Hardware System," *Fuzzy Sets and Systems*, 32, 1989, pp. 161–180.

CHAPTER 8

1. Takashi Ono, "Eveything's Fuzzy Among Household Appliances," *The Japan Times Weekly International Edition*, August 27–September 2, 1990, p. 13.

2. "Japan: Will It Lose Its Competitive Edge?" *Business Week*, April 27, 1992, pp. 51–54, p. 51.

3. "Mitsubishi to Market Car That Uses 'Fuzzy' Logic," Associated Press, April 6, 1992.

CHAPTER 9

1. Judea Pearl, "The Bayesian Approach," in Glenn Shafer and Judea Pearl, eds., *Readings in Uncertain Reasoning*, San Mateo, CA, Morgan Kaufmann Publishers, 1990, pp. 339–344, p. 342.

2. Judea Pearl, "Bayesian and Belief-Function Formalisms for Evidential Reasoning: A Conceptual Analysis," in ibid., pp. 540–574, p. 541.

3. In Theodore M. Porter, *The Rise of Statistical Thinking, 1820–1900*, Princeton, NJ, Princeton University Press, 1986, p. 75.

4. E. T. Jaynes, "Bayesian Methods: General Background," in James H. Justice, ed., *Maximum Entropy and Bayesian Methods in Applied Statistics*, Cambridge, UK, Cambridge University Press, 1986, pp. 1–25, p. 8.

5. Bradley Efron, "Why Isn't Everyone a Bayesian?" (originally written in 1986), in Shafer and Pearl, *Uncertain Reasoning*, pp. 21–25.

6. Dennis Lindley, comment on Efron, ibid., p. 27.

7. E. T. Jaynes, "Bayesian Methods," p. 17.

8. John Fox, "Eclecticism in the Parish of Uncertainty," *The Knowledge Engineering Review*, 3, 1988, pp. 85–86, p. 85.

9. Peter Cheeseman, "In Defense of 'An Inquiry into Computer Understanding,'" *Computer Intelligence*, 4, 1988, pp. 129–142, p. 136.

10. Peter Cheeseman, "Probabilistic versus Fuzzy Reasoning," in L. N. Kapal and J. F. Lemmer, eds., *Uncertainty in Artificial Intelligence*, New York, Elsevier Science Publishers B.F. (North-Holland), 1986, pp. 85–102, p. 88.

11. Dennis Lindley, "The Probability Approach to the Treatment of Uncertainty in Artificial Intelligence and Expert Systems," *Statistical Science*, 2(1), 1987, pp. 17–29, p. 17.

12. Ibid., p. 21.

13. Ibid., p. 22.

14. George Klir, "Is There More to Uncertainty Than Some Probability Theorists Might Have Us Believe?" *International Journal of General Systems*, 15, 1989, pp. 347–378, p. 360.

15. Glenn Shafer, comment on Dennis Lindley, "The Probability Approach," p. 37.

16. John Lemmer, "A System Perspective for Choosing Between Dempster-Shafer and Bayes," *The Knowledge Engineering Review*, 3, 1988, pp. 74–78, p. 76.

17. Stephen Watson, comment on Dennis Lindley, "The Probability Approach," p. 32.

18. A. P. Dempster and Augustine Kong, comment on Dennis Lindley, "The Probability Approach," p. 33.

19. John Fox, "Parish of Uncertainty," p. 85.

20. Hesiod, *Works and Days*, in *Hesiod and Theognis*, trans. Dorothea Wender, New York, Penguin Books, 1973, p. 81.

CHAPTER 10

Much of the theoretical material in this chapter comes from Bart Kosko, *Neural Networks and Fuzzy Systems*, Englewood Cliffs, NJ, Prentice-Hall, 1991, chapters 1 and 7.

1. Lotfi Zadeh, "Toward a Theory of Fuzzy Systems," in R. E. Kalman and N. DeClaris, eds., *Aspects of Network and System Theory*, New York, Holt, Rinehart & Winston, 1971, p. 486.

2. George J. Klir and Tina A. Folger, *Fuzzy Sets*, pp. 108–109.

3. Dennis Lindley, response to commenters, "The Probability Approach," p. 42.

4. E. T. Jaynes, "Bayesian Methods," p. 19.

5. Technically, it is the intersection of a set and its opposite divided by the union of the set and its opposite.

6. Bart Kosko, "Fuzzy Systems as Universal Approximators," unpublished paper, 1992, pp. 16–17. The quotes in this section, however, come from personal discussions.

CHAPTER 11

1. Friedrich Nietzsche, *The Gay Science*, §354, in *A Nietzsche Reader*, trans. R. J. Hollingdale, New York, Penguin Books, 1977, p. 66.

2. Reviewer's memo to Robert Balzer, November 3, 1980.

3. Nils Nilsson, "On *Logical Foundations of Artificial Intelligence*, a Response to the Reviews by S. Smoliar and J. Sowa," *AI*, February 1989, p. 133.

4. In M. Mitchell Waldrop, *Man-Made Minds: The Promise of Artificial Intelligence*, New York, Walker & Co., 1987, p. 36.

5. Ibid., p. 39.

6. William A. Taylor, *What Every Engineer Should Know About Artificial Intelligence*, Cambridge, MA, MIT Press, 1988, p. 160.

7. In Elisabetta Binaghi, "A Fuzzy Logic Inference Model for a Rule-Based System in Medical Diagnosis," *Expert Systems*, 7, August 1990, pp. 134–141, p. 136.

8. Glenn Shafer, comment on Dennis Lindley, "The Probability Approach," p. 38.

9. Ian Graham, "Fuzzy Logic in Commercial Expert Systems—Results and Prospects," *Fuzzy Sets and Systems*, 40, 1991, pp. 451–472, p. 456.

10. Lotfi Zadeh, "The Role of Fuzzy Logic in the Management of Uncertainty in Expert Systems," *Fuzzy Sets and Systems*, 11, 1983, pp. 199–227, p. 203.

11. Lotfi Zadeh, "Analysis of Complex Systems and Decision Processes," p. 40.

12. Jun-ichiroh Fujimoto, Tomofumi Nakatuni, and Masahide Yoneyama, "Speaker-Independent Word Recognition Using Fuzzy Pattern Matching," *Fuzzy Sets and Systems*, 32, 1989, pp. 181–191.

13. Gregg Oden and Dominic Massaro, "Integration of Featural Information in Speech Perception," *Psychological Review*, 85, 1978, pp. 172–191.

CHAPTER 12

1. Stephen M. Kosslyn and Oliver Koenig, *Wet Mind: The New Cognitive Neuroscience*, New York, Free Press, 1992, p. 42.
2. For a fuller description, see Bart Kosko, *Neural Networks and Fuzzy Systems*, chapter 11.
3. Atsushi Morita and Akio Noda, "Fuzzy Model of Neural Network Type," in *Proceedings of the 1990 National Convention IEEE Japan—Industry Applications Society*, pp. I.13–I.20.
4. *Proceedings of the International Fuzzy Engineering Symposium '91, November 13–15, Yokohama, Japan, Fuzzy Engineering toward Human Friendly Systems*, vol. 1, Tokyo, Ohmsha, 1991, pp. 515–561.

CHAPTER 13

1. Sheridan Tatsuno, *Created in Japan: From Imitators to World-Class Innovators*, New York, Harper & Row, 1990.

CHAPTER 14

1. Simon Loe, "SGS-Thompson Gets into Fuzzy Logic," *Electronic Buyers' News*, August 12, 1991, pp. 10, 12.
2. In early 1992 Zadeh suggested that Yamakawa place a mouse on the platform and let the animal dart back and forth on it. Zadeh calls this problem "more interesting" than the triple pendulum. Yamakawa tried it, and the pole still stayed upright. "In Japan this has produced a sensation," says Zadeh, who adds that neither neural nets nor classic control can achieve this feat.

BIBLIOGRAPHY

■

BOOKS

Alletzhauser, Albert J. *The House of Nomura*. New York: Little, Brown and Co., 1990.

Aristotle. *Categories* and *De Interpretatione* (J. L. Ackrill, trans.). Oxford, UK: Oxford University Press, 1963.

———. *De Anima* (Hugh Lawson-Tancred, trans.). New York: Penguin Books, 1986.

———. *The Metaphysics* (Hippocrates G. Apostle, trans.). Bloomington: Indiana University Press, 1966.

———. *The Nicomachean Ethics* (J. E. C. Welldon, trans.). London: Macmillan and Co., 1892.

Austin, J. L. *Sense and Sensibilia*. Oxford, UK: Clarendon Press, 1962.

Beniger, James R. *The Control Revolution*. Cambridge, MA: Harvard University Press, 1986.

Berlin, Brent, and Paul Kay. *Basic Color Terms: Their Universality and Evolution*. Berkeley: University of California Press, 1969.

Bhattacharyya, Narendra Nath. *Jain Philosophy: Historical Outline*. New Delhi: Munshiram Manoharlal Publishers Pvt. Ltd., 1976.

Burchfield, Robert. *The English Language*. Oxford, UK: Oxford University Press, 1986.

Cajori, Florian. *A History of Mathematics*. New York: Macmillan Co., 1961.

Campbell, Jeremy. *The Improbable Machine*. New York: Simon & Schuster, 1989.

Darwin, Charles. *On the Origin of Species*. London: Murray, 1859.

Dauben, Joseph W. *Georg Cantor: His Mathematics and Philosophy of the Infinite*. Cambridge, MA: Harvard University Press, 1979.

Drexler, K. Eric. *Engines of Creation: The Coming Era of Nanotechnology*. New York: Anchor Press, 1986.

Dumitriu, Anton. *History of Logic*, vol. 4. Tunbridge Wells, UK: Abacus Press, 1977.

Gardner, Howard. *The Mind's New Science: A History of the Cognitive Revolution*. New York: Basic Books, 1985.

Gardner, Martin. *Logic Machines and Diagrams*. Chicago: University of Chicago Press, 1958.

Geertz, Clifford. *Local Knowledge: Further Essays in Interpretive Anthropology*. New York: Basic Books, 1983.

Gergen, Kenneth J. *The Saturated Self*. New York: Basic Books, 1991.

Gjertsen, Derek. *Science and Philosophy: Past and Present*. New York: Penguin Books, 1989.

Graubard, Stephen R., ed. *The Artificial Intelligence Debate*. Cambridge, MA: MIT Press, 1988.

Grayeff, Felix. *Aristotle and His School*. London: Duckworth, 1974.

Guthrie, W. K. C. *A History of Greek Philosophy*, vol. 6, *Aristotle: An Encounter*. Cambridge, UK: Cambridge University Press, 1981.

Hodges, Wilfrid. *Logic*. New York: Penguin Books, 1977.

Hume, David. *A Treatise of Human Nature*. New York: Penguin Books, 1969.

Johnson, George. *In the Palaces of Memory*. New York: Alfred A. Knopf, 1991.

Jorrand, P., and V. Sgurev, eds. *Artificial Intelligence*, vol. 4, *Methodology, Systems, and Applications*. Amsterdam: North-Holland, 1990.

Kalghatgi, T. G. *Jaina Logic*. New Delhi, India: Shri Raj Krishen Jain Charitable Trust, 1984.

Kamioka, Kazuyoshi. *Japanese Business Pioneers*. Union City, CA: Heian International, 1988.

Kaufmann, Arnold. *Introduction to the Theory of Fuzzy Subsets*. New York: Academic Press, 1975.

Kearns, Robert L. *Zaibatsu America*. New York: Free Press, 1992.

Kent, Beverley. *Charles S. Peirce: Logic and the Classification of the Sciences*. Montreal: McGill–Queen's University Press, 1987.

Kline, Morris. *Mathematics: The Loss of Certainty*. New York: Oxford University Press, 1980.

———. *Mathematics in Western Culture*. New York: Oxford University Press, 1953.

Klir, George J., and Tina A. Folger. *Fuzzy Sets, Uncertainty, and Information*. Englewood Cliffs, NJ: Prentice-Hall, 1988.

Kosko, Bart. *Neural Networks and Fuzzy Systems*. Englewood Cliffs, NJ: Prentice-Hall, 1991.

Kosslyn, Stephen M., and Oliver Koenig. *Wet Mind: The New Cognitive Neuroscience*. New York: Free Press, 1992.

Kuhn, Thomas. *The Structure of Scientific Revolutions*. Chicago: University of Chicago Press, 1962.

Lakoff, George. *Women, Fire, and Dangerous Things: What Categories Reveal about the Mind*. Chicago: University of Chicago Press, 1987.

Lao Zi. *Tao Te Ching* (Victor H. Mair, trans.). New York: Bantam Books, 1990.

Locke, John. *An Essay Concerning Human Understanding*. London: Bye and Law, 1805.

Łukasiewicz, Jan. *Selected Works* (L. Borkowski, ed.). London: North-Holland, 1970.

Mackie, J. L. *Truth, Probability and Paradox*. Oxford, UK: Clarendon Press, 1973.

Massaro, Dominic. *Speech Perception by Eye and Ear: A Paradigm for Psychological Inquiry*. Hillsdale, NJ: Lawrence Erlbaum Associates, 1987.

Mayr, Ernst. *The Growth of Biological Thought*. Cambridge, MA: Belknap Press, 1982.

McCall, Storrs, ed. *Polish Logic, 1920–1939*. Oxford, UK: Oxford University Press, 1967.

Mill, John Stuart. *Selected Writings of John Stuart Mill*. New York: New American Library, 1968.

Minsky, Marvin, and Seymour Papert. *Perceptrons*. Cambridge, MA: MIT Press, 1969.

Moore, Charles, ed. *The Japanese Mind: Essentials of Japanese Philosophy and Culture*. Honolulu: University of Hawaii Press, 1987.

Nagel, Ernest, and James R. Newman. *Gödel's Proof*. New York: New York University Press, 1958.

Needham, Joseph. *Science and Civilization in China*. Cambridge, UK: Cambridge University Press, 1954.

Negoitia, C. W., and Daniel Ralescu. *Applications of Fuzzy Sets to Systems Analysis*. Basel and Stuttgart: Birkhäuser, 1975.

Nietzsche, Friedrich. *The Gay Science* (Walter Kaufmann, trans.). New York: Random House, 1974.

———. *A Nietzsche Reader* (R. J. Hollingdale, trans.). New York: Penguin, 1977.

Novák, Vilém. *Fuzzy Sets and Their Applications.* Philadelphia; Adam Hilger, 1989.

Pagels, Heinz R. *The Dreams of Reason.* New York: Bantam Books, 1989.

Peirce, Charles Sanders. *Collected Papers of Charles Sanders Peirce* (Charles Hartshorne and Paul Weiss, eds.), vols. 5 and 6. Cambridge, MA: Harvard University Press, 1934, 1935.

Piovesana, Gino K. *Contemporary Japanese Philosophical Thought.* New York: St. John's University Press, 1969.

Porter, Theodore M. *The Rise of Statistical Thinking, 1820–1900.* Princeton, NJ: Princeton University Press, 1986.

Prestowitz, Clyde. *Trading Places: How We Allowed Japan to Take the Lead.* New York: Basic Books, 1988.

Quine, W. V. *Philosophy of Logic.* Cambridge, MA: Harvard University Press, 1970.

———. *Quiddities: An Intermittently Philosophical Dictionary.* Cambridge, MA: Belknap Press, 1987.

Rescher, Nicholas. *Many-valued Logic.* New York: McGraw-Hill, 1969.

Riasanovsky, Nicholas. *A History of Russia.* New York: Oxford University Press, 1984.

Rosenblatt, Frank. *Principles of Neurodynamics.* New York: Spartan Books, 1963.

Ross, David. *Aristotle.* New York: Methuen & Co., 1985.

Russell, Bertrand. *A History of Western Philosophy.* New York: Simon & Schuster, 1945.

Sansom, G. B. *Japan: A Short Cultural History.* New York: D. Appleton, 1931.

Scholz, Heinrich. *Concise History of Logic.* New York: Philosophical Library, 1961.

Shafer, Glenn, and Judea Pearl, eds. *Readings in Uncertain Reasoning.* San Mateo, CA: Morgan Kaufmann Publishers, 1990.

Shodt, Frederik L. *Inside the Robot Kingdom.* Tokyo: Kodansha International, 1988.

Smithson, Michael. *Ignorance and Uncertainty: Emerging Paradigms.* New York: Springer-Verlag, 1989.

Stcherbatsky, F. T. *Buddhist Logic,* vol. 1. New York: Dover Publications, 1962.

Sugeno, Michio, ed. *Industrial Applications of Fuzzy Control.* Amsterdam: North-Holland, 1985.

Szaniawski, Klemens, ed. *The Vienna Circle and the Lvov-Warsaw School.* Boston: Kluwer Academic Publishers, 1989.

Tatsuno, Sheridan. *Created in Japan: From Imitators to World-Class Innovators.* New York: Harper & Row, 1990.

Taylor, William A. *What Every Engineer Should Know About Artificial Intelligence.* Cambridge, MA: MIT Press, 1988.

Toulmin, Stephen. *Cosmopolis: The Hidden Agenda of Modernity.* New York: Free Press, 1990.

Varley, H. Paul. *Japanese Culture.* Honolulu: University of Hawaii Press, 1973.

Waldrop, M. Mitchell. *Man-Made Minds: The Promise of Artificial Intelligence.* New York: Walker & Co., 1987.

Wing-tsit Chan, ed. and trans. *A Source Book in Chinese Philosophy.* Princeton, NJ: Princeton University Press, 1963.

Woleński, Jan. *Logic and Philosophy in the Lvov-Warsaw School.* Dordrecht, The Netherlands: Kluwer Academic Publishers, 1989.

Wu, Laurence C. *Fundamentals of Chinese Philosophy.* New York: University Press of America, 1986.

Zemankova-Leech, Maria, and Abraham Kandel. *Fuzzy Relational Data Bases: A Key to Expert Systems.* Cologne: Verlag TÜV Rheinland, 1984.

Zerubavel, Eviatar. *The Fine Line: Making Distinctions in Everyday Life.* New York: Free Press, 1991.

Zimmermann, Hans-Jürgen. *Fuzzy Set Theory and Its Applications, Second Edition.* Boston: Kluwer Academic Publishers, 1991.

ARTICLES

Agnew, Neil, and John Brown. "Foundations for a Model of Knowing. I. Constructing Reality," *Canadian Psychology,* 30, 1989, pp. 152–67.

———. "Foundations for a Model of Knowing. II. Fallible but Functional Knowledge," *Canadian Psychology,* 30, 1989, pp. 168–83.

Armstrong, Robert. "Why 'Fuzzy Logic' Beats Black-or-White Thinking," *Business Week,* May 21, 1990, pp. 92–93.

Ayyub, Bilal M. "Systems Framework for Fuzzy Sets in Civil Engineering," *Fuzzy Sets and Systems,* 40, 1991, pp. 491–508.

Bellman, Richard, R. Kalaba, and Lotfi Zadeh. "Abstraction and Pattern Classification," *Journal of Mathematical Analysis and Applications,* 13(1), January 1966, pp. 1–7.

Bellman, Richard, and Lotfi Zadeh. "Decision-Making in a Fuzzy Environment," *Management Science,* 17, 1970, pp. B141–64.

Benjamin, A. Cornelius. "Science and Vagueness," *Philosophy of Science,* 6, 1939, pp. 422–31.

Berliner, Hans. "Computer Backgammon," *Scientific American,* 242(6), June 1980, pp. 54–60.

Binaghi, Elisabetta. "A Fuzzy Logic Inference Model for a Rule-Based System in Medical Diagnosis," *Expert Systems,* 7, August 1990, pp. 134–41.

Black, Max. "Reasoning with Loose Concepts," *Dialogue,* 2, 1963, pp. 1–12.

———. "Vagueness: An Exercise in Logical Analysis," *Philosophy of Science,* 4, 1937, pp. 427–55.

Bochvar, Dmitri A. "On a Three-valued Logical Calculus and Its Application to the Analysis of the Paradoxes of the Classical Extended Functional Calculus" (orig. 1937), Merrie Bergmann, trans., *History and Philosophy of Logic,* 2, 1981, pp. 87–112.

Brody, Herb. "The Neural Computer," *Technology Review,* August/September 1990, pp. 43–49.

Burgess, Don, Willett Kempton, and Robert MacLaury. "Tarahumara Color Modifiers: Category Structure Presaging Evolutionary Change," *American Ethnologist,* 10, 1983, pp. 133–49.

Chameau, J. L. A., A. Alteschaeffl, H. L. Michael, and J. T. P. Yao. "Potential Applications of Fuzzy Sets in Civil Engineering," *International Journal of Man-Machine Studies,* 19, 1983, pp. 9–18.

Cheeseman, Peter. "In Defense of 'An Inquiry into Computer Understanding,'" *Computer Intelligence,* 4, 1988, pp. 129–42.

———. "Probabilistic versus Fuzzy Reasoning." In L. N. Kanal and J. F. Lemmer, eds., *Uncertainty in Artificial Intelligence.* New York: Elsevier Science Publishers B.F. (North-Holland), 1986, pp. 85–102.

Chen Shi-quan. "Fuzzy Classification of Acute Toxicity of Poisons," *Fuzzy Sets and Systems,* 27, 1988, pp. 255–60.

Cooper, Leon. "A Possible Organization of Animal Memory and Learning." In James A. Anderson and Edward Rosenfeld, eds., *Neurocomputing: Foundations of Research.* Cambridge, MA: MIT Press, 1988, pp. 195–207.

Cox, R. T. "Probability, Frequency, and Reasonable Expectation," *American Journal of Physics,* 17, 1946, pp. 1–13.

Da Rocha, Armando Freitas. "Neural Fuzzy Point Processes," *Fuzzy Sets and Systems*, 5, 1981, pp. 127–40.

Efron, Bradley. "Why Isn't Everyone a Bayesian?" In Glenn Shafer and Judea Pearl, eds., *Readings in Uncertain Reasoning*. San Mateo, CA: Morgan Kaufmann Publishers, 1990, pp. 21–25.

Efron, Sonni. "A Cultural Trade Imbalance," *Los Angeles Times*, May 21, 1992, pp. 1, A10–A11.

Ezard, John. "Trying to Understand a Pigeon's English," *The Guardian*, January 9, 1979.

Fox, John. "Eclecticism in the Parish of Uncertainty," *The Knowledge Engineering Review*, 3, 1988, pp. 85–86.

Frankel, Eugene. "Corpuscular Optics and the Wave Theory of Light: The Science and Politics of a Revolution in Physics," *Social Studies of Science*, 6, 1976, pp. 141–84.

Fujimoto, Jun-ichiroh, Tomofumi Nakatuni, and Masahide Yoneyama. "Speaker-Independent Word Recognition Using Fuzzy Pattern Matching," *Fuzzy Sets and Systems*, 32, 1989, pp. 181–91.

Gaines, Brian. "Foundations of Fuzzy Reasoning," *International Journal of Man-Machine Studies*, 8, 1976, pp. 623–68.

———. "Logical Foundations for Database Systems." In Ebrahim Mamdani and Brian Gaines, eds., *Fuzzy Reasoning and Its Applications*. New York: Academic Press, 1981, pp. 289–308.

———. "Precise Past—Fuzzy Future," *International Journal of Man-Machine Studies*, 19, 1983, pp. 117–34.

Gordon, Jean, and Edward H. Shortliffe, "The Dempster-Shafer Theory of Evidence." In Glenn Shafer and Judea Pearl, eds., *Readings in Uncertain Reasoning*. San Mateo, CA: Morgan Kaufmann Publishers, 1990, pp. 529–39.

Graham, Ian. "Fuzzy Logic in Commercial Expert Systems—Results and Prospects," *Fuzzy Sets and Systems*, 40, 1991, pp. 451–72.

Grattan-Guinness, Ivor. "The Manuscripts of Emil L. Post," *History and Philosophy of Logic*, 11, 1990, pp. 77–83.

Haack, Susan. "Do We Need Fuzzy Logic?" *International Journal of Man-Machine Studies*, 11, 1979, pp. 437–45.

Hakel, Milton D. "How Often Is Often?" *American Psychologist*, 23, 1968, pp. 533–34.

Hempel, Carl. "Vagueness and Logic," *Philosophy of Science*, 6, 1939, pp. 163–80.

Hersh, Harry, and Alfonso Caramazza."A Fuzzy Set Approach to Modifiers and Vagueness in Natural Language," *Journal of Experimental Psychology: General*, 105, 1976, pp. 254–76.

Hirota, Kaoru, Yoshinori Arai, and Shiroh Hachisu. "Fuzzy Controlled Robot Arm Playing Two-Dimensional Ping-Pong Game," *Fuzzy Sets and Systems*, 32, 1989, pp. 149–59.

Hirota, Kaoru, and Kazuhiro Ozawa. "Fuzzy Flip-Flop and Fuzzy Registers," *Fuzzy Sets and Systems*, 32, 1989, pp. 139–48.

Holmblad, Peter, and Jens-Jorgen Østergaard. "Control of a Cement Kiln by Fuzzy Logic." In M. M. Gupta and Elie Sanchez, eds., *Fuzzy Information and Decision Processes*. Amsterdam: North-Holland, 1982, pp. 398–99.

"Japan: Will It Lose Its Competitive Edge?" *Business Week*, April 27, 1992, pp. 51–54.

Jaynes, E. T. "Bayesian Methods: General Background." In James H. Justice, ed., *Maximum Entropy and Bayesian Methods in Applied Statistics*. Cambridge, UK: Cambridge University Press, 1986, pp. 1–25.

Jordan, Zbigniew A. "The Development of Mathematical Logic in Poland Between the Two Wars." In Storrs McCall, ed., *Polish Logic, 1920–1939*. Oxford, UK: Oxford University Press, 1967, pp. 346–97.

Kay, Paul, and Chad McDaniel. "The Linguistic Significance of the Meanings of Basic Color Terms," *Language*, 54, 1978, pp. 610–46.

Keller, James, Deeka Subhangkasen, and Kenneth Unklesbay. "An Approximate Reasoning Technique for Recognition in Color Images of Beef Steaks," *International Journal of General Systems*, 16, 1990, pp. 331–42.

Kempton, Willett. "Category Grading and Taxonomic Relations: A Mug Is a Sort of Cup," *American Ethnologist*, 5, 1978, pp. 44–65.

Klir, George. "Is There More to Uncertainty Than Some Probability Theorists Might Have Us Believe?" *International Journal of General Systems*, 15, 1989, pp. 347–78.

———. "Measures and Principles of Uncertainty and Information: Recent Developments." In H. Atmanspacher and H. Scheingraber, eds., *Information Dynamics*. New York: Plenum Press, 1991, pp. 1–14.

———. "A Principle of Uncertainty and Information Invariance," *International Journal of General Systems*, 17, 1991, pp. 249–75.

Klir, George, and Arthur Ramer. "Uncertainty in the Dempster-Shafer Theory: A Critical Re-examination," *International Journal of General Systems*, 18, 1990, pp. 155–66.

Kosko, Bart. "Fuzzy Entropy and Conditioning," *Information Sciences*, 40, 1986, pp. 165–74.

Kuwahara, Hiroshi, Mituo Harada, Yasuhide Seno, and Koji Kinoshita. "Application of Fuzzy Theory to the Control of Shield Tunneling." In *The Proceedings of Third IFSA Congress*, August 6–11, 1989, pp. 250–53.

Labov, William. "The Boundaries of Words and Their Meanings." In Charles-James N. Bailey and Roger W. Shuy, eds., *New Ways of Analyzing Variation in English*. Washington, DC: Georgetown University Press, 1972, pp. 340–73.

Lakoff, George. "Hedges: A Study in Meaning Criteria and the Logic of Fuzzy Concepts," *Journal of Philosophical Logic*, 2, 1973, pp. 458–508.

Larsen, P. Martin. "Industrial Applications of Fuzzy Logic Control." In Ebrahim Mamdani and Brian Gaines, eds., *Fuzzy Reasoning and Its Applications*. New York: Academic Press, 1981, pp. 335–42.

Lea, Robert N. "Automated Space Vehicle Control for Rendezvous Proximity Operations," *Telematics and Informatics*, 5(3), 1988, pp. 179–85.

Le Blond, "Aristotle on Definition." In Jonathan Barnes, Malcolm Schofield, and Richard Sorabji, eds., *Articles on Aristotle*, vol. 3. *Metaphysics*. New York: St. Martin's Press, 1979, pp. 63–79.

Leibniz, Gottfried W. "Of Universal Synthesis and Analysis," in G.R.N. Parkinson, ed., *Gottfried Wilhelm Leibniz: Philosophical Writings*. London: J.M. Dent & Sons, 1973.

Lemmer, John. "A System Perspective for Choosing Between Dempster-Shafer and Bayes," *The Knowledge Engineering Review*, 3, 1988, pp. 74–78.

Li Zuoyang. "A Model of Weather Forecast by Fuzzy Grade Statistics," *Fuzzy Sets and Systems*, 26, 1988, pp. 275–81.

Lindley, Dennis V. "The Present Position in Bayesian Statistics," *Statistical Science*, 5(1), 1990, pp. 44–89.

———. "The Probability Approach to the Treatment of Uncertainty in Artificial Intelligence and Expert Systems," *Statistical Science*, 2(1), 1987, pp. 17–29.

Loe, Simon. "SGS-Thompson Gets into Fuzzy Logic," *Electronic Buyers' News*, August 12, 1991, pp. 10, 12.

Łukasiewicz, Jan. "Aristotle on the Law of Contradiction." In Jonathan Barnes,

Malcolm Schofield, and Richard Sorabji, eds., *Articles on Aristotle*, vol. 3, *Metaphysics*. New York: St. Martin's Press, 1979, pp. 50–62.

Mamdani, Ebrahim, and Sedrak Assilian. "An Experiment in Linguistic Synthesis with a Fuzzy Logic Controller," *International Journal of Man-Machine Studies*, 7, 1975, pp. 1–13.

Massaro, Dominic. "Language Processing and Information Integration." In Norman H. Anderson, ed., *Information Integration Theory*, vol. I, *Cognition*. Hillsdale, NJ: Lawrence Erlbaum Associates, 1991, pp. 259–92.

————. "Testing Between the TRACE Model and the Fuzzy Logical Model of Speech Perception," *Cognitive Psychology*, 21, 1989, pp. 398–421.

Massaro, Dominic, and Daniel Friedman. "Models of Integration Given Multiple Sources of Information," *Psychological Review*, 97, 1990, pp. 225–52.

Maydole, Robert. "Paradoxes and Many-Valued Set Theory," *Journal of Philosophical Logic*, 4, 1975, pp. 269–91.

McCorduck, Pamela. "Artificial Intelligence: An Aperçu." In Stephen R. Graubard, ed., *The Artificial Intelligence Debate*. Cambridge, MA: MIT Press, 1988, pp. 65–83.

McGill, V. J. "Concerning the Laws of Contradiction and Excluded Middle," *Philosophy of Science*, 6, 1939, pp. 196–211.

Meisels, Amnon, Abraham Kandel, and Gay Gecht. "Entropy and the Recognition of Fuzzy Letters," *Fuzzy Sets and Systems*, 31, 1989, pp. 297–309.

Munson, John. "As the Sun Sets . . . : Concerning Fuzziness and a Search for the Real World in Hawaii," *Computer Group News*, March 1968, pp. 20–21.

Murakami, S., F. Takemoto, H. Fujimura, and E. Ide. "Weld-Line Tracking Control of Arc-Welding Robot Using Fuzzy Logic Controller," *Fuzzy Sets and Systems*, 32, 1989, pp. 221–37.

Murayama, Y., T. Terano, S. Masui, and N. Akiyama. "Optimizing Control of a Diesel Engine." In Michio Sugeno, ed., *Industrial Applications of Fuzzy Control*. Amsterdam: North-Holland, 1985, pp. 63–71.

Nadin, Mihai. "The Logic of Vagueness." In Eugene Freeman, ed., *The Relevance of Charles Peirce*. La Salle, IL: Monist Library of Philosophy, 1983, pp. 154–66.

Newstead, Stephen, and Richard Griggs. "Fuzzy Quantifiers as an Explanation of Set Inclusion Performance," *Psychological Research*, 46, 1984, pp. 377–88.

Nishikawa, Taizo. "Fuzzy Theory: The Science of Human Intuition," *Japan Computer Quarterly*, 79, 1989, pp. 25–37.

O'Brien, Stephen. "The Ancestry of the Giant Panda," *Scientific American*, 257, November 1987, pp. 102–7.

Oden, Gregg. "Fuzziness in Semantic Memory: Choosing Exemplars of Subjective Categories," *Memory & Cognition*, 5, 1977, pp. 198–204.

————. "A Fuzzy Logical Model of Letter Identification," *Journal of Experimental Psychology*, 5, 1979, pp. 336–52.

————. "Integration of Fuzzy Linguistic Information in Language Comprehension," *Fuzzy Sets and Systems*, 14, 1984, pp. 29–41.

Oden, Gregg, and Dominic Massaro. "Integration of Featural Information in Speech Perception," *Psychological Review*, 85, 1978, pp. 172–91.

Ono, H., T. Ohnishi, and Y. Terada. "Combustion Control of Refuse Incineration Plant by Fuzzy Logic," *Fuzzy Sets and Systems*, 32, 1989, pp. 192–206.

Ono, Takashi. "Eveything's Fuzzy Among Household Appliances," *The Japan Times Weekly International Edition*, August 27–September 2, 1990, p. 13.

Osherson, Daniel, and Edward Smith. "Gradedness and Conceptual Combination," *Cognition*, 12, 1982, pp. 299–318.

————. "On the Adequacy of Prototype Theory as a Theory of Concepts," *Cognition*, 9, 1981, pp. 35–58.

Pal, Sankar K. "Fuzzy Tools for the Management of Uncertainty in Pattern Recognition, Image Analysis, Vision and Expert Systems," *International Journal of Systems Science*, 22, 1991, pp. 511–49.

Pearl, Judea. "Bayesian and Belief-Function Formalisms for Evidential Reasoning: A Conceptual Analysis." In Glenn Shafer and Judea Pearl, eds., *Readings in Uncertain Reasoning*. San Mateo, CA: Morgan Kaufmann Publishers, 1990, pp. 540–74.

Pfeilsticker, Arne. "The Systems Approach and Fuzzy Set Theory: Bridging the Gap Between Mathematical and Language-Oriented Economists," *Fuzzy Sets and Systems*, 6, 1981, pp. 209–33.

Ponsard, Claude. "Fuzzy Mathematical Models in Economics," *Fuzzy Sets and Systems*, 28, 1989, pp. 273–83.

Rogers, Michael. "The Future Looks 'Fuzzy,' " *Newsweek*, May 26, 1990, pp. 46–47.

Rosch, Eleanor. "Natural Categories," *Cognitive Psychology*, 4, 1973, pp. 328–50.

———. "On the Internal Structure of Perceptual and Semantic Categories." In Timothy Moore, ed., *Cognitive Development and the Acquisition of Language*. New York: Academic Press, 1973, pp. 111–44.

———. "Principles of Categorization." In Eleanor Rosch and Barbara Lloyd, eds., *Cognition and Categorization*. Hillsdale, NJ: Lawrence Erlbaum Associates, 1978, pp. 27–48.

Rosch, Eleanor, and Carolyn B. Mervis. "Family Resemblances: Studies in the Internal Structure of Categories," *Cognitive Psychology*, 7, 1975, pp. 573–605.

Rosenblatt, Frank. "The Perceptron: A Probabilistic Model for Information Storage and Organization in the Brain," *Psychological Review*, 65, 1958, pp. 386–408.

Ruspini, Enrique. "Fuzzy Logic in the Flakey Robot." In *The Proceedings of the International Conference on Fuzzy Logic and Neural Networks*, Iizuka, Japan, July 20–24, 1990, pp. 767–70.

Russell, Bertrand. "Vagueness," *Australasian Journal of Psychology and Philosophy*, 1, 1923, pp. 84–92.

Sasaki, Tsuna, and Takamasa Akiyama, "Traffic Control Process of Expressway by Fuzzy Logic," *Fuzzy Sets and Systems*, 26, 1988, pp. 165–78.

Shafer, Glenn. "Probability Judgment in Artificial Intelligence and Expert Systems," *Statistical Science*, 2(1), 1987, pp. 3–16.

Shaw-Kwei, Moh. "Logical Paradoxes for Many-Valued Systems," *Journal of Symbolic Logic*, 19(1), 1954, pp. 37–40.

Simpson, Ray. "The Specific Meanings of Certain Terms Indicating Differing Degrees of Frequency," *The Quarterly Journal of Speech*, 30, 1944, pp. 328–30.

Smets, Philippe. "Medical Diagnosis: Fuzzy Sets and Degrees of Belief," *Fuzzy Sets and Systems*, 5, 1981, pp. 259–66.

Smithson, Michael. "Fuzzy Set Theory and the Social Sciences: The Scope for Applications," *Fuzzy Sets and Systems*, 26, 1988, pp. 1–21.

Sobocinski, Boleslaw. "In Memoriam Jan Łukasiewicz," *Philosophical Studies (Ireland)*, 6, 1956, pp. 3–49.

Sugeno, Michio, T. Murofushi, T. Mori, T. Tatematsu, and J. Tanaka, "Fuzzy Algorithmic Control of a Model Car by Oral Instructions," *Fuzzy Sets and Systems*, 32, 1989, pp. 207–19.

Sugeno, Michio, and M. Nishida. "Fuzzy Control of Model Car," *Fuzzy Sets and Systems*, 1985, 165, pp. 103–13.

Sugeno, Michio, O. Yagishita, and O. Itoh. "Application of Fuzzy Reasoning to the Water Purification Process." In Michio Sugeno, ed., *Industrial Applications of Fuzzy Control*. Amsterdam: North-Holland, 1985, pp. 19–39.

Takagi, Tomohiro, and Michio Sugeno. "Derivation of Fuzzy Control Rules from Human Operators' Control Actions," in Elie Sanchez, ed., *Fuzzy Information*,

Knowledge Representation, and Decision Analysis. Oxford, UK: Pergamon Press, 1984.

Togai, Masaki, and Hiroyuke Watanabe. "A VLSI Implementation of a Fuzzy-Inference Engine: Toward an Expert System on a Chip," *Information Sciences,* 38, 1986, pp. 147–63.

Tversky, Amos, and Daniel Kahneman. "Probability, Representativeness, and the Conjunction Fallacy," *Psychological Review,* 90, 1983, pp. 293–315.

Ujihara, Hideyo, and Shintaro Tsuji. "The Revolutionary AI-2100 Elevator-Group Control System and the New Intelligent Option Series," *Mitsubishi Electric Advance,* December 1988, pp. 5–8.

Umbers, I. G., and P. J. King. "An Analysis of Human Decision-Making in Cement Kiln Control and the Implications for Automation." In Ebrahim Mamdani and Brian Gaines, eds., *Fuzzy Reasoning and Its Applications.* New York: Academic Press, 1981, pp. 369–81.

Wagner, Wolfgang. "A Fuzzy Model of Concept Representation in Memory," *Fuzzy Sets and Systems,* 6, 1981, pp. 11–26.

Wang Peizhuang, Liu Xihui, and Elie Sanchez. "Set-Valued Statistics and Its Application to Earthquake Engineering," *Fuzzy Sets and Systems,* 18, 1986, pp. 347–56.

Weaver, Warren. "Science and Complexity," *American Scientist,* 36, 1948, pp. 536–44.

Yamakawa, Takeshi. "Stabilization of an Inverted Pendulum by a High-Speed Fuzzy Logic Controller Hardware System," *Fuzzy Sets and Systems,* 32, 1989, pp. 161–80.

Yasunobu, Chizuko, Tetsuya Maruoka, Kazuhide Shigemi, and Makoto Shimazaki. "A Knowledge-Based Technical Analysis System for Financial Decision-Making." In *The Proceedings of the Pacific Rim International Conference on Artificial Intelligence '90,* Nagoya, Japan, November 14–16, 1990, pp. 89–94.

Yasunobu, Seiji, and T. Hasegawa. "Evaluation of an Automatic Container Crane Operation System Based on Predictive Fuzzy Control," *Control-Theory and Advanced Technology,* 2, 1986, pp. 419–32.

Yasunobu, Seiji, and Shoji Miyamoto. "Automatic Train Operation System by Predictive Fuzzy Control." In Michio Sugeno, ed., *Industrial Applications of Fuzzy Control.* Amsterdam: North-Holland, 1985, pp. 1–18.

Zadeh, Lotfi. "Biological Application of the Theory of Fuzzy Sets and Systems." In *The Proceedings of an International Symposium on Biocybernetics of the Central Nervous System.* Boston: Little, Brown, 1969, pp. 199–206.

———. "The Concept of a Linguistic Variable and Its Application to Approximate Reasoning—I," *Information Sciences,* 8, 1975, pp. 199–249.

———. "The Concept of a Linguistic Variable and Its Application to Approximate Reasoning—II," *Information Sciences,* 8, 1975, pp. 301–57.

———. "The Concept of a Linguistic Variable and Its Application to Approximate Reasoning—III," *Information Sciences,* 9, 1975, pp. 43–80.

———. "From Circuit Theory to System Theory," *Proceedings of Institution of Radio Engineers,* 50, 1962, pp. 856–65.

———. "Fuzzy Logic and Approximate Reasoning," *Synthese,* 30, 1975, pp. 407–28.

———. "Fuzzy Sets," *Information and Control,* 8(3), June 1965, pp. 338–53.

———. "Fuzzy Sets as a Basis for a Theory of Possibility," *Fuzzy Sets and Systems,* 1, 1978, pp. 3–28.

———. "Fuzzy Sets versus Probability," *Proceedings of the IEEE,* 68(3), March 1980, p. 421.

———. "Making Computers Think Like People," *IEEE Spectrum,* August 1984, pp. 26–32.

———. "A Note on Prototype Theory and Fuzzy Sets," *Cognition*, 12, 1982, pp. 291–97.

———. "Outline of a New Approach to the Analysis of Complex Systems and Decision Processes," *IEEE Transactions on Systems, Man, and Cybernetics*, SMC-3(1), January 1973, pp. 28–44.

———. "Probability Measures of Fuzzy Events," *Journal of Mathematical Analysis and Applications*, 23(2), August 1968, pp. 421–27.

———. "A Rationale for Fuzzy Control," *Journal of Dynamic Systems, Measurement, and Control*, 94, March 1972, pp. 3–4.

———. "The Role of Fuzzy Logic in the Management of Uncertainty in Expert Systems," *Fuzzy Sets and Systems*, 11, 1983, pp. 199–227.

———. "A System-Theoretic View of Behavior Modification." In Harvey Wheeler, ed., *Beyond the Punitive Society*. San Francisco: W. H. Freeman, 1973, pp. 160–70.

———. "Thinking Machines: A New Field in Electrical Engineering," *Columbia Engineering Quarterly*, January 1950, pp. 12–13, 30–31.

———. "Toward Fuzziness in Computer Systems: Fuzzy Algorithms and Languages." In G. Boulaye, ed., *Architecture and Design of Digital Computers*. Paris: Dunod, 1971, pp. 9–18.

———. "Toward a Theory of Fuzzy Systems." In R. E. Kalman and N. DeClaris, eds., *Aspects of Network and System Theory*. New York: Holt, Rinehart & Winston, 1971, pp. 469–90.

Zelený, Milan. "Cybernetyka," *International Journal of General Systems*, 13, 1987, pp. 289–94.

INDEX

ABOUT THE AUTHORS

■

Daniel McNeill has written numerous books and articles on high-tech, and his work has also appeared in fiction, travel, history, law, and education publications. He is a graduate of Harvard Law School and lives with his wife in Los Angeles, California.

Paul Freiberger has written about advanced technology for more than a decade. He coauthored *Fire in the Valley: The Making of the Personal Computer*, and covered Silicon Valley for six years at the *San Francisco Examiner*. He now works at Interval Research Corp., which is pursuing the high-tech breakthroughs of the 1990s and beyond, and lives in San Mateo, California, with his wife and nineteen-month-old son.